Springer Theses

Recognizing Outstanding Ph.D. Research

Aims and Scope

The series "Springer Theses" brings together a selection of the very best Ph.D. theses from around the world and across the physical sciences. Nominated and endorsed by two recognized specialists, each published volume has been selected for its scientific excellence and the high impact of its contents for the pertinent field of research. For greater accessibility to non-specialists, the published versions include an extended introduction, as well as a foreword by the student's supervisor explaining the special relevance of the work for the field. As a whole, the series will provide a valuable resource both for newcomers to the research fields described, and for other scientists seeking detailed background information on special questions. Finally, it provides an accredited documentation of the valuable contributions made by today's younger generation of scientists.

Theses may be nominated for publication in this series by heads of department at internationally leading universities or institutes and should fulfill all of the following criteria

- They must be written in good English.
- The topic should fall within the confines of Chemistry, Physics, Earth Sciences, Engineering and related interdisciplinary fields such as Materials, Nanoscience, Chemical Engineering, Complex Systems and Biophysics.
- The work reported in the thesis must represent a significant scientific advance.
- If the thesis includes previously published material, permission to reproduce this must be gained from the respective copyright holder (a maximum 30% of the thesis should be a verbatim reproduction from the author's previous publications).
- They must have been examined and passed during the 12 months prior to nomination.
- Each thesis should include a foreword by the supervisor outlining the significance of its content.
- The theses should have a clearly defined structure including an introduction accessible to new PhD students and scientists not expert in the relevant field.

Indexed by zbMATH.

More information about this series at https://link.springer.com/bookseries/8790

Andrei-Constantin Braitor

Advanced Hierarchical Control and Stability Analysis of DC Microgrids

Doctoral Thesis accepted by
The University of Sheffield, Sheffield, UK

Springer

Author
Dr. Andrei-Constantin Braitor
Department of Automatic Control
and Systems Engineering
University of Sheffield
Sheffield, UK

Supervisor
Prof. George C. Konstantopoulos
Department of Electrical and Computer
Engineering
University of Patras
Rion, Greece

ISSN 2190-5053 ISSN 2190-5061 (electronic)
Springer Theses
ISBN 978-3-030-95414-7 ISBN 978-3-030-95415-4 (eBook)
https://doi.org/10.1007/978-3-030-95415-4

This Springer imprint is published by the registered company Springer Nature Switzerland AG
The registered company address is: Gewerbestrasse 11, 6330 Cham, Switzerland

This thesis is dedicated to my parents,

Vicky and Adrian

Supervisor's Foreword

The rapid increase of renewable energy systems utilisation has rendered power systems as one of the most promising areas of research in the last decades. The large integration of distributed energy resources (DERs), i.e. renewable energy sources, energy storage systems and responsive loads, into the electricity grid currently represents an essential requirement for every country in order to meet its energy targets. Every power distribution system is, therefore, being transformed towards a 'decentralised grid' consisting of multiple small-scale power systems integrated together or operating autonomously (islanded), where each one of them consists of distributed generation units, energy storage components and local residential and industrial loads, forming the so-called 'microgrids'. Among the different types of microgrids (AC, DC, hybrid), the concept of 'DC microgrids' has gained particular attention as it represents a localised solution for utilising DER units that already operate on DC power, e.g. photovoltaic systems, batteries, fuel cells, led lights, etc., without the need of converting their power to AC, which leads to reduced energy efficiency.

The stable, reliable and optimal operation of DC microgrids mainly depends on the control mechanisms of the DERs, which are responsible for the power delivered to the microgrid and the regulation of their local DC voltage close to the rated/desired value. This is often accomplished through multi-layered control that follows a hierarchical structure, where primary control is used to guarantee the stability of the microgrid and provide ancillary services, while secondary control is required to achieve accurate power sharing among multiple DERs and restoration of a local DC bus voltage to the rated value. For the appropriate control design and the stable operation of the DC microgrid, the accurate dynamic model of the microgrid that includes the power converter units for the DERs and the nonlinear loads is, therefore, essential.

In the present thesis, Dr. Andrei-Constantin Braitor investigates the nonlinear dynamic description of different types of DC microgrids that includes the dynamics of the power converter devices (DC/DC and AC/DC converters) and nonlinear loads (constant power loads) and then proposes novel control schemes for the primary and secondary control layers. The main aim of the proposed control algorithms is to maintain the already well accepted by the industry control concepts (e.g. droop control) but implement them in a particular structure that accomplishes two tasks: (i) maintain a desired bound for each DER's injected current/voltage according to their operating limits and (ii) guarantee stability for the entire microgrid system. Dr. Braitor utilises fundamental nonlinear systems theory to achieve the desired current/voltage limitation and such an approach consequently facilitates the rigorous proof of the closed-loop system stability. Particularly, opposed to conventional approaches that investigate stability on a particular DC microgrid example, in this thesis, Dr. Braitor presents a new framework for obtaining analytic conditions for the microgrid stability, even with the presence of nonlinear loads, without relying on a given set of system parameters. Hence, this approach provides a very useful tool for DC microgrid system operators and control designers regarding the appropriate selection of the system and control parameters. The proposed control methods have been extensively tested in both simulation environments and experimental testbeds, and the results of this research have been published in highly ranked international journals and conferences.

As a Supervisor, I would like to thank Dr. Braitor for his patience, motivation and continuous effort to deliver high-level research in such a challenging and important subject. It has been a pleasure to work with him all these years and see him grow as an independent researcher. I believe that Dr. Braitor's thesis can act as an inspiration for future researchers who would like to work in the areas of control and power systems, since it demonstrates how fundamental control and system theory can be combined with power system research and provide novel solutions to real-life problems, such as the stable and reliable operation of modern microgrids.

Patras, Greece
December 2021

Dr. George C. Konstantopoulos

Abstract

The paradigm shift in electrical power grids and the increased interest towards decentralisation have opened a new window in the design, control and theoretical analysis of small-scale power systems, i.e. microgrids, which aim at the integration and utilisation of renewable energy sources, energy storage systems and responsive loads at a local scale. Given their DC nature, DC microgrids have attracted significant interest as they provide a natural interface to the main grid by avoiding additional conversion steps. The aim of this thesis is to design and analyse novel hierarchical control schemes, both at the primary and secondary control level, that guarantee tight voltage regulation, accurate power sharing, current or voltage limitation, and present a straightforward approach towards deriving stability conditions for DC microgrids that incorporate nonlinear loads.

As the microgrid configuration is paramount in the theoretical analysis, a rigorous method of computing the admittance matrix is developed that facilitates the stability analysis of DC microgrid systems supplying a constant impedance (Z), constant current (I) or constant power (P) load. This method is particularly useful as it permits the factorisation of the admittance matrix, while separating the singular matrices. In this way, the closed-loop asymptotic stability proof can be more easily approached by isolating the singularities and then, employing straightforward linear algebra tools, such as quadratic eigenvalue problem (QEP) theory, to derive the stability conditions.

As stated in the technical specifications of every source, a crucial issue is represented by the ability to protect itself and its interface device (power converter) during faults, transients and unrealistic power demands. That is why current-limiting control is often combined with the primary control, i.e. often droop control, in DC microgrids to ensure the desired unit protection. In this thesis, novel current-limiting droop controllers are formulated for DC microgrids consisting initially of multiple unidirectional DC/DC boost converters and, later on, bidirectional DC/DC boost and three-phase AC/DC converters, that integrate different distributed generation units with the local grid. In contrast to the traditional approaches that use small-signal modelling, here, the accurate nonlinear models of the converter units are taken into account to prove the boundedness and the current-limiting property, by employing Lyapunov methods and the ultimate boundedness theory. Exponential stability at the desired equilibrium point is mathematically guaranteed and further analysed from a graphical perspective, providing insights into the load's effect on the system performance and stability.

To address the shortcomings that conventional droop control approaches introduce, i.e. inaccurate power sharing and significant load voltage drop, a DC microgrid architecture that takes into account the power converter dynamics of distributed generation units is deployed under decentralised primary and distributed secondary control scheme, in a hierarchical control framework. At the primary control layer, a novel current-limiting droop control scheme is implemented in a decentralised manner, whereas at the secondary control layer, a fully distributed controller that performs a voltage restoration and improves the power sharing is deployed using a nearest-neighbour coupling communication network. By investigating for the first time both the dynamics of the converters with the nonlinear load and the two-layer control, singular perturbation theory is applied to analytically guarantee the stability of the entire DC microgrid.

Finally, apart from the desired overcurrent protection, since DC capacitors are customarily used at the output of each converter unit to stabilise the output voltage, they also introduce a maximum voltage limit. Hence, as the need for protection against overvoltages has emerged, droop controllers with inherent overvoltage protection are also proposed for parallel and meshed configuration networks. The upper limit of the voltage of each source is rigorously proven using the ultimate boundedness theory. Asymptotic stability to the desired equilibrium for the closed-loop system is analytically guaranteed, and detailed conditions are derived to guide the control design.

In order to validate the effectiveness of the different control methodologies developed in this thesis, both simulation and experimental testing are performed for each one of the methods and are also compared with the conventional approaches to highlight their superiority.

Acknowledgements

Reflecting on my years spent at The University of Sheffield, I often ask myself *Would I do it again?*. It seemed like four years flew by in a second. I reckon time does fly by when you are enjoying what you are doing.

My deepest gratitude goes to my supervisor, Dr. George C. Konstantopoulos, for his amazing assistance and mentoring skills, patient guidance, enthusiastic encouragement and useful critiques. His determination and drive towards supporting, advising and guiding me throughout my entire Ph.D. journey has had a profound impact on my research work. And his willingness to help extended beyond the theoretical academic challenges and eventually led to more hands-on endeavours such as ACSE Remote Lab and ADEPT projects.

I would like to offer my thanks to all my colleagues in the office Alex Paspatis, Seyfullah Dedeoglu, Pablo Baldivieso-Monasterios, GregorisMichos and Panagiotis Skenzos. The assistance provided by Dr. Shlomo Gadelovits when conducting lab work has been of great value in testing and validating controllers experimentally. Equally useful were the people from group meetings and journal clubs, my colleague Fernando Genis-Mendoza and several other guests that occasionally came to visit. The support and assistance I received from all of you are highly appreciated.

Furthermore, I would like to express my gratitude to my collaborators, first on the hybrid-electric aircraft paper and later on the Co-design Microgrid project, namely Hadriano Morales-Escamilla, Dr. Andy Mills at the UTC Rolls Royce and, of course, Prof. Visakan Kadirkamanathan, my second supervisor. Thank you for a good experience. I have learned a lot from our gatherings together.

Last but not least, I would like to thank my girlfriend, Diana, for her patience and understanding, my friends and family for constantly trying to keep me away from the office and lab and making me to postpone and delay work, such as writing this thesis. I guess, in the end, you failed guys... it's finished.

And, the answer to the question I started with is: *Definitely, yes!*.

Contents

List of Figures

List of Tables

Chapter 1
Introduction

In this opening chapter the general idea of the thesis is presented. The early explanation aims towards justifying the rationale behind pursuing such a research topic. Following a compact introduction of the challenges, a complete set of aims and objectives are being predetermined to provide a clear distinction regarding where this research work fits into the academic literature. To identify and support the general idea, the main contributions are plainly stated, followed by a brief presentation of the structure of the thesis.

1.1 General Topic and Overview

With the beginning of the current century, a series of multiple emerging trends came to light that threaten to affect the environment, health and quality of life. Energy demand worldwide is projected to increase by 53% from 2008 to 2035, with the highest growth by 2.3% recorded last year, according to the International Energy Agency (IEA) [1], its fastest pace this decade, an exceptional performance driven by a robust global economy. As several problems have emerged associated with environment and fossil fuel depletion, green energy resources have attracted a lot of attention in the past years, causing the demand for renewable energy to skyrocket.

To cope with the increasing energy demand and environmental regulation, research is pushed towards incorporating the renewable sources into the grid, an initiative that led to the rush of finding new hybrid architectures to transition from the vast centralised traditional grid to interconnected smaller scale grids, coined *microgrids* (MGs). The MG paradigm, proposed as a standard reference of the future grid, was first introduced in [2] as a *"cluster of loads and microsources operating as a single controllable system that provides both power and heat"*. A more complete definition of a microgrid is given in [3], where microgrids are defined as *"electricity distribution systems containing loads and distributed energy sources, such as dis-*

A.-C. Braitor, *Advanced Hierarchical Control and Stability Analysis of DC Microgrids*, Springer Theses, https://doi.org/10.1007/978-3-030-95415-4_1

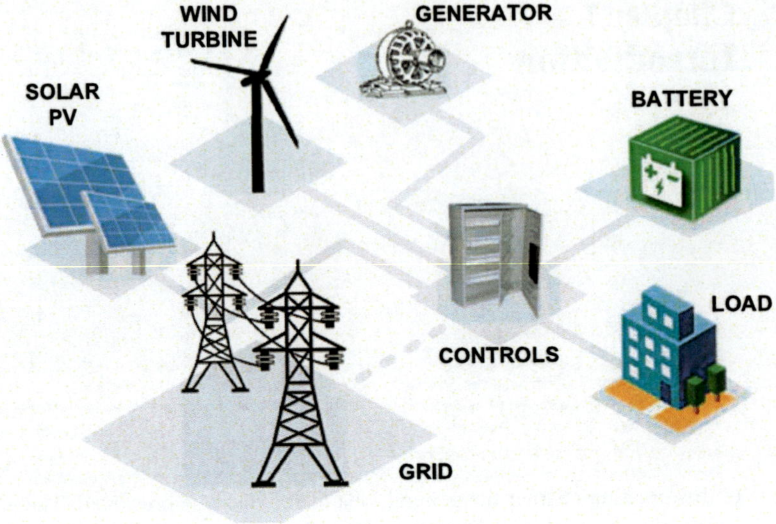

Fig. 1.1 Microgrid concept diagram

tributed generators, storage devices, or controllable loads, that can be operated in a controlled, coordinated way either while connected to the main power network, in grid-connected mode, or while islanded, if they are isolated from the main grid". Figure 1.1 conceptually shows the main elements available in a microgrid.

Different from the conventional AC frameworks, DC microgrids are able to improve the power quality, minimise power losses, decrease energy conversion steps and operating costs, thus, boosting the advantages and value of distributed energy resources (DER). Widely regarded as a capable technology to ensure efficient coupling between renewable energy sources (RES), energy storage systems (ESS) and DC loads, DC microgrids have gained a significant increase in focus and interest [4, 5]. That is mainly since DC systems provide a higher efficiency and reliability, a simpler control structure and expandability, as well as a natural interface to the increased numbers of DC renewable generations, storage systems and electronic loads [6]. Moreover, as the power now has a DC nature, one can avoid the synchronisation issues and instability effects that follow with it.

Various applications (e.g. electronic devices, batteries or photovoltaic panels) can be directly connected to the DC bus, reducing the multi AC/DC conversion stages and, thus, avoiding unstable scenarios and mitigating power conversion losses introduced by the frequency and reactive power control. The advantages rendered by DC microgrids have been confirmed in a wide range of industry applications, such as electric vehicles (EVs) [7], electric trains [8], (more-, hybrid- or all-) electric aircrafts [9, 10], and electric ships [11]. The benefits of using a DC ship power system have been summarised in [12], as follows:

- improvement of prime mover efficiency and fuel costs reduction;
- weight and space savings;
- generators operating with a unity power factor;
- lower transmission losses;
- faster and simpler parallel connection of generators; and
- simpler implementation of energy storage.

But apart from the development of industrial, commercial or residential DC distribution networks, several promising DC framework applications can be found in EV charging stations or smart buildings [13].

However, a pivotal role in the integration of the RES and ESS is played by the centrepiece of these frameworks that interfaces these units to the microgrid system, namely the power converter. Both the single-phase DC/DC or three-phase AC/DC power converters are often controlled using pulse-width-modulation (PWM) techniques to guarantee output voltage regulation, bidirectional power flow, low harmonic distortion of the grid current and unity power factor. In DC network architectures, power converter units are generally connected either in cascaded, meshed or, as in most cases, in a parallel configuration to a common bus supplying common loads. Particularly in parallel configurations, the main challenges one needs to address are related to: (i) constant terminal bus voltage; (ii) accurate load power sharing between parallel sources; (iii) power quality; (iv) fault protection; and (v) system instability [14]. Hence, the control of the converter units within the DC microgrid is crucial for providing a stable and reliable DC network architecture, and it is mainly designed in a hierarchical control structure [15].

The standard primary control approach, located at the lower level of the multi-level control and responsible for the stability of the microgrid, usually rests on droop-based methods [16]. Still, droop control strategies incorporate an output virtual resistance at each source placing it further away from the ideal (current/voltage) source, resulting in the local DC bus voltage being more dependent on the load coupled to the system. Such strategies are utilised in microgrids where communication networks are not suitable for data exchange due to distributed physical locations of the distributed generation sources, with the condition that the line impedances are mainly inductive. The conventional droop approach presents some serious shortcomings, i.e. poor voltage regulation due to their permanent offset, inaccuracies in the load power sharing, dependency on line impedance, slow dynamic response, and low stability margin. Despite the emergence of other droop method variations that claim to tackle these drawbacks, the latter stability aspect is of major significance and, in most cases, insufficiently addressed.

Ensuring a system is stable is always a daunting task due to the complex dynamics that the power converters and the nonlinear loads introduce. A generic model for loads, coined "ZIP", has been been adopted in [17] and incorporates constant impedance load (CIL) or Z load, constant current load (CCL) or I load, and the most challenging to deal with, constant power load (CPL) or P load. Under tight-speed regulation, the motor drive, similar to tight regulated downstream power converters, manifests constant power behaviour at the DC bus and creates unique dynamic characteristics. That is why it has been an active research topic for years [18]. Being one of the most common, constant power loads are prone to yield instability at the DC bus, which propagates within the entire system, as they exhibit a negative impedance characteristic in small-signal analysis, which overcomplicates the stabilisation problem. This destabilising effect is often referred to as *negative impedance instability* [19]. However, to guarantee system stability when having CPLs, in small-signal analysis, one must ensure that the impedance inequality criteria is satisfied.

Passive damping methods are one type of approach to shape the output impedance of the input filter by adding an extra physical resistance, but this causes additional power losses. In addressing this issue, active damping methods, based on cascaded systems without input filters between two-stage converters were introduced. Although popular in small-scale microgrids, these strategies are not applicable in relatively large DC microgrids. A significant deal of research works (see [20] and references within) looked into the stability and stabilisation problem of single or multiple converter droop-controlled DC microgrids with CPLs, leading to the conclusion that system tends to be unstable when traditional decentralised or distributed control is implemented independently. Reduced order models have been proposed to obtain stable ranges of droop coefficients and safe operating regions, but the implemented assumptions and limitations proved too unrealistic, as it involves knowledge of system and loads parameters, which is not always easy to come by, and possible only in fictional scenarios.

Hierarchical control strategies are often adopted not only to guarantee system stability, but also to boost the system performance. Moreover, these strategies introduce a certain degree of independence between their stacked control layers and increase the reliability of the system as they continue to operate even when failures occur in one of the upper control layers. Multi-level control is attained by simultaneously utilising local converter and digital communication link-based coordinated control, which are separated by at least one order of magnitude in control bandwidth. With the development of the communication technology, the multi-level distributed control becomes more and more popular in DC microgrids. Distributed control methods, employing a diffusive (or nearest neighbour coupling) [21, 22] have been proposed in the literature, with the main upside that the system maintains its full functionality, even in case of communication failure of some of the links, given the network remains connected. That is because the distributed control is immune to single point of failure (SPOF). Nevertheless, the stability of the microgrid system, particularly feeding CPLs, under multi-level distributed control has not been adequately studied, mainly due to the complex dynamics introduced by the system, nonlinear loads and multi-level controller. The vast majority of the stability techniques that investigate DC networks are based on converter small-signal models and linear approximations, utilising the Middlebrook and Cuk criterion [23]. While this strategy is convenient as one obtains the gain of the open-loop system, by taking into account the loads' input impedance and the sources' output impedance, the stability results that are often obtained are based on the parameters of a given case study and cannot be generalised. Hence, studying the stability of the overall DC microgrid system under hierarchical control still remains an open problem.

Aside from the theoretical analysis and stability proof, another crucial feature arises from the devices technical specifications, namely their capability to protect oneself during faults, transients and unrealistic load power demands. Existing protection methods rely on hardware units, such as utilising additional fuses, circuit breakers and/or protection relays [24–26]. But still, designing controllers with a current-limiting feature remains a challenge [27, 28]. Current-limiting strategies based on saturated PI controllers are commonly employed to limit and fix an upper bound for the current, however the limitations have not been properly overcome, e.g.

(i) saturated PI controllers limit only the reference of the input current, hence, during transients the overcurrent protection is often violated [29];
(ii) one can not guarantee closed-loop stability as the controller can suffer from integrator wind-up that may yield instability in the system [30].

But overcurrents are not the only unwanted phenomenon, as the need of protecting the equipment against overvoltages is equally important [31, 32]. Overvoltage instances happen when the voltage in the circuit, or part of the circuit, increases above its designed limit causing potential damage in the equipment of the grid. As DC capacitors are customary utilised at the output of each converter unit to stabilise the output voltage, their presence establishes a maximum voltage limitation. A great deal of upper layer controllers aimed to mitigate the voltage rise impact by means of optimal power flow, wavelet packet decomposition (WPD) and general regression neural networks (GRNN) theory or by reducing the active power injected by a source until its local voltage complies with the operational requirements, namely active power curtailment (APC). Implementing an overvoltage protection at the primary control level still constitutes a challenge yet to be addressed.

In the following sections, the aims and objectives of this thesis are presented. A list of contributions is also included, followed by a summary with the structure of this work.

1.2 Motivation and Scope

The previous subsection has highlighted the advantages that DC microgrid systems display, indicated the possible loads one can find in DC systems, and stressed the importance of having a good controller to ensure the desired, and necessarily stable operation. As already mentioned, system stability is of paramount importance, and a non-trivial task to guarantee analytically. Furthermore, the possibility of guaranteeing microgrid protection through control has also been identified as an active research work direction. But, an approach that merges all the above aspects and considers the described DC microgrid scenario has never been explored.

In this thesis, a combined control approach is proposed for converter units in DC microgrids that integrates several features such as tight voltage regulation, accurate load power distribution, current-limiting capability or overvoltage protection. Moreover, the theoretical stability analysis is also developed to ensure analytically that the entire system is stable. This perspective aims to cover several gaps and solve several problems still present in the literature.

1.2.1 Challenges

Control design and theoretical analysis of DC microgrids is far from straightforward, particularly when the interfacing units are represented by converters with nonlinear dynamics, structured in various network topologies and feeding nonlinear loads. And aside from the theoretical development, simulation and experimental testing also represent a key part, and bear a significant weight in the validation process.

The main challenges one faces when working on a project of this type can be summed up in the following points:

- develop a method that expresses analytically the network topology of the units, that might ease the theoretical analysis and facilitate the acquisition of stability criteria;
- design a control framework that integrates the following characteristics:
  - keep the interfacing devices output voltage constant and tightly regulated to the desired level of the terminal DC voltage bus;
  - ensure uninterruptible power supply of the load during faults or abnormal conditions, such as unrealistic power demands;
  - maintain the power quality and avoid the overusage of the sources by implementing an accurate power distribution proportional with the devices ratings;
  - protect the equipment at all times during transients, such as overcurrents, or overvoltages, caused by the sudden connection or disconnection of a component in the network;
- a major challenge is to analytically guarantee closed-loop system stability for the entire system, which incorporates the converter network with the nonlinear loads, and the controller;
- testing a controller in a DC microgrid framework, both experimentally or numerically, requires time. That is mainly because it involves getting familiar with the pieces of software and/or hardware, and because it requires time for troubleshooting the emerging issues.

1.2.2 Aims and Objectives

As a result, in this thesis one tries to meet the challenges identified above. The purpose of this work revolves around addressing current open problems in the literature. The main aims and objectives are included below:

1. Elaborate an analytic strategy that computes and factorises the Laplacian matrix of the units network in order to permit a straightforward path towards investigating the overall system stability.
2. Implement a unified primary controller that can incorporate aforementioned tasks, namely:
   (a) tight output voltage regulation;
   (b) accurate load power distribution among sources;
   (c) an inherent current/voltage limitation capability.
3. Develop a secondary controller (in a hierarchical framework) to improve the performance of the lower layers.
4. Investigate stability of the entire system incorporating both the plant and the hierarchical control architecture.
5. Verify the developed control methods by performing simulation and experimental testing.

1.2.3 Contribution

The major contributions of this work are as follows:

- **Nonlinear system investigation**: Unlike most approaches that consider mainly buck converters which have linear dynamics, the present work gives a particular attention to unidirectional/bidirectional DC/DC boost converters and bidirectional three-phase AC/DC rectifiers which are nonlinear systems.
- **Admittance matrix**: A different analytic approach is presented as a key tool for computing and suitably factorising the admittance matrix of a DC microgrid configuration, consisting of n sources and a common constant impedance, current or power load, that enables and facilitates the acquisition of stability conditions.
- **Closed-loop system models**: Compared to the generic static droop control strategy, in here, dynamic droop control frameworks are proposed for each unit, which also includes the measurement of the bus voltage (in parallel configuration network) that further complicates the analysis.
- **Stability analysis**: By employing theories such as standard (SEP), generalised (GEP), or quadratic eigenvalue problem (QEP) and other linear matrix analysis tools to the closed-loop system and utilising the factorisation of the admittance matrix, stability conditions are obtained to guide the control design.
- **Droop control strategy superiority**: Appropriate comparisons are being made to highlight the improved power sharing and tighter voltage regulation achieved by the proposed controllers.
- **Microgrid protection**: Inherent current or voltage limitation is guaranteed by the proposed control designs for every converter unit at the primary layer, unlike existing strategies that aim to implement microgrid protection at the upper control layers.

- **Numerical and experimental validation**: The proposed approaches are then verified through both simulation and, in most cases, experimental testing. To highlight the benefits of the present approaches, comparisons with existing methods are also included wherever possible.

1.2.4 List of Publications

1. A.-C. Braitor, G.C. Konstantopoulos and V. Kadirkamanathan, "Power sharing of parallel operated DC-DC converters using current-limiting droop control", 2017 25h Mediterranean Conference on Control and Automation (MED), Valletta, 2017, pp. 528–533, https://doi.org/10.1109/MED.2017.7984171.
2. A.-C. Braitor, P.R. Baldivieso-Monasterios, G.C. Konstantopoulos and V. Kadirkamanathan, "Current-limiting droop control design of paralleled AC-DC and DC-DC converters in DC micro-grids", IECON 2018 - 44th Annual Conference of the IEEE Industrial Electronics Society, Washington DC, 2018, pp. 132–137, https://doi.org/10.1109/IECON.2018.8591990.
3. A.-C. Braitor, A. R. Mills, V. Kadirkamanathan, G.C. Konstantopoulos, P.J. Norman and C.E. Jones, "Control of DC power distribution system of a hybrid electric aircraft with inherent overcurrent protection", 2018 IEEE International Conference on Electrical Systems for Aircraft, Railway, Ship Propulsion and Road Vehicles and International Transportation Electrification Conference (ESARS-ITEC), Nottingham, 2018, pp. 1–6, https://doi.org/10.1109/ESARS-ITEC.2018.8607779.
4. A.-C. Braitor, G.C. Konstantopoulos and V. Kadirkamanathan, "Current-limiting droop control design and stability analysis for paralleled boost converters in DC microgrids", in IEEE Transactions on Control Systems Technology, https://doi.org/10.1109/TCST.2019.2951092.
5. A.-C. Braitor, G.C. Konstantopoulos and V. Kadirkamanathan, "Stability analysis and nonlinear current-limiting control design for DC microgrids with CPLs" in IET Smart Grid, https://doi.org/10.1049/IET-STG.2019.0235.
6. A.-C. Braitor, G.C. Konstantopoulos and V. Kadirkamanathan, "Admittance matrix computation and stability analysis of droop controlled DC micro-grids with CPL", IFAC PapersOnLine, Volume 53, Issue 2, 2020, pp. 13525–13530, https://doi.org/10.1016/j.ifacol.2020.12.769.
7. A.-C. Braitor, G.C. Konstantopoulos and V. Kadirkamanathan, "Novel droop control design for overvoltage protection of DC microgrids with a constant power load", 2020 28th Mediterranean Conference on Control and Automation (MED), 2020, pp. 875–880, https://doi.org/10.1109/MED48518.2020.9182891.
8. A.-C. Braitor, G.C. Konstantopoulos and V. Kadirkamanathan, "Enhanced primary droop controller for meshed DC microgrids with overvoltage protection" 2021 29th Mediterranean Conference on Control and Automation (MED), 2021, pp. 368–373, https://doi.org/10.1109/MED51440.2021.9480294.

9. A.-C. Braitor, G.C. Konstantopoulos and V. Kadirkamanathan, "Stability analysis of DC microgrids with CPL under novel decentralised primary and distributed secondary control", Automatica, 2022 [accepted].

1.3 Thesis Outline

The remainder of this thesis is organised in the following way.

Chapter 2 presents a thorough literature review of the current modelling, control and stability analysis strategies and approaches employed in DC microgrids. It revises the schematics and mathematical models of most commonly used power converters in DC microgrids, and also explains the models of ordinary loads one can expect to find. This chapter reviews several hierarchical control strategies that aim to guarantee the desired microgrid operation. An important attention is given to stability analysis of droop-controlled, and/or hierarchical controlled DC microgrids. Another important aspect touched in this chapter is the DC microgrid protection against overcurrents and overvoltages. In the end, a detailed list of limitations and assumptions is taken out, to potentially identify research gaps where further improvement can be done.

In Chap. 3, common notations are introduced that are used throughout the entire thesis. For one to follow and understand the theoretical analysis, several important theoretical tools are revisited. This chapter presents elementary graph theory notions, and establishes straightforward linear matrix analysis tools such as Sylvester's law of inertia or the quadratic eigenvalue problem, as they are paramount for understanding the developed stability theory. Ultimate boundedness and singular perturbation theory are also discussed as they occupy a key role in designing the controllers and employing the time scale separation prior to investigating the system stability.

Chapter 4 proposes an alternative way of computing and factorising the admittance matrix for parallel operated converters in DC microgrids feeding Z, I or P loads. The proposed methodology enables and facilitates the acquisition of stability conditions particularly when the controller measures the output voltage (voltage regulation, droop control etc.) since it permits the factorisation of the impedance matrix by separating singular matrices. As a result, closed-loop stability proof becomes trivial with the isolation of singular matrices, and uses simple linear algebra tools to obtain stability conditions. To validate the proposed approach, compute the admittance matrix and test the stability conditions, a DC microgrid with n DC/DC converters supplying a Z, I or P load is considered. To demonstrate the desired operation of the DC microgrid control and design framework, simulation tests are diligently carried out.

As overcurrent protection guaranteed through control became of interest, a nonlinear current-limiting droop controller is presented in Chap. 5, for paralleled DC/DC unidirectional boost converters supplying a common load. In Sect. 5.1, the robust droop control method is used and implemented within the system as a dynamic virtual resistance in series with the converter inductor. The proposed framework achieves voltage regulation and power sharing, with an inherent current limitation for all converters regardless of the load type and magnitude variations. In Sect. 5.2, with an improved version of the proposed controller in Sect. 5.1, which does not need the measurement of the currents at the output, sufficient stability conditions are analytically derived, using singular perturbation theory and linear matrix analysis tools, for the overall microgrid system having either a Z, an I, or a P load. A comparison with conventional droop control methods is also provided to highlight the proposed controller superiority. Both simulation and experimental testing are being performed with a CPL, subject to several changes of the constant power.

In Chap. 6, a DC microgrid is considered consisting of bidirectional three-phase AC/DC rectifiers and DC/DC boost converters supplying a common load. The proposed approach introduces two controllers, in Sect. 6.1, to ensure voltage regulation, reactive power control and load power sharing with an inherent current-limiting capability, independently from the system or load parameters. This concept is based on the idea of incorporating a constant virtual resistance and a bounded dynamic virtual controllable voltage which can be both both positive and negative leading to two-way power flow. Unlike previous section, where the converters feed a CIL, in Sect. 6.2, for stability analysis purposes, the CPL case is investigated. For the closed-loop system stability with the developed current-limiting controller, sufficient conditions are mathematically derived considering separate operating scenarios. The stability conditions are also graphically illustrated, and they provide valuable insights of the CPL's influence onto the system performance and stability. The developed controllers are validated by comparison with the conventional PI control strategy. Also, experimental results are included to test the effectiveness of the proposed control method on a real experimental setup.

Chapter 7 investigates the stability of a DC microgrid feeding a CPL under a decentralised primary and a distributed secondary control scheme. At the primary control layer, the droop expression is suitably formulated and implemented using the recently proposed state-limiting PI controller to accomplish an inherent current limitation for each converter and simultaneously facilitate the theoretical analysis. Acquiring limited information from the common bus and neighbouring converters, a distributed secondary controller is introduced to enhance the power sharing and tighten the voltage regulation. Applying the singular perturbation theory allows, for the first time, to analyse both the dynamics of the converters and the CPL together with the hierarchical control scheme, providing the possibility of guaranteeing analytically the stability of the entire DC microgrid system. To support the theoretical findings, simulation and experimental tests are being performed.

Chapter 8 proposed an overvoltage protection for both parallel and mesh configuration of DC microgrids. In Sect. 8.1, a droop controller for parallel operated converters in DC microgrids with a CPL is proposed to achieve desired load voltage regulation and accurate load power distribution, whilst ensuring an overvoltage protection for each source. The upper limit of the voltage of each source is rigorously proven using ultimate boundedness theory. Analytic conditions are also derived to guide the control design. In Sect. 8.2, converters in a mesh configuration are investigated, all of which are feeding a local CPL. The droop-based methodology inherits the conventional droop control features and additionally guarantees an overvoltage protection capability of each source. Asymptotic stability to the desired equilibrium for the closed-loop system is mathematically proven, and sufficient conditions are given to conduct the control design implementation. Simulation testings are carried out, in both sections, to validate the theoretical contributions and the advantages of the newly developed droop controllers.

In Chap. 9, a brief summary of the entire work is put into place. Several conclusions and final thoughts are drawn. Some closing considerations are mentioned on the assumptions made throughout the whole thesis. In the end, this thesis provides a few ideas on the future directions of this work and potential next steps.

References

1. IEA. International energy agency (2020), http://www.iea.org
2. R.H. Lasseter, Microgrids, in *2002 IEEE Power Engineering Society Winter Meeting. Conference Proceedings (Cat. No.02CH37309)*, vol. 1 (IEEE, New York, 2002), pp. 305–308
3. C. Marnay, S. Chatzivasileiadis, C. Abbey, R. Iravani, G. Joos, P. Lombardi, P. Mancarella, J. Von Appen, Microgrid evolution roadmap, in *2015 International Symposium on Smart Electric Distribution Systems and Technologies (EDST)*, pp. 139–144 (2015)
4. T. Dragičević, X. Lu, J.C. Vasquez, J.M. Guerrero, DC microgrids—part i: a review of control strategies and stabilization techniques. IEEE Trans. Power Electron. **31**(7), 4876–4891 (2016)
5. T. Dragičević, X. Lu, J.C. Vasquez, J.M. Guerrero, DC microgrids—part ii: a review of power architectures, applications, and standardization issues. IEEE Trans. Power Electron. **31**(5), 3528–3549 (2016)
6. M. Rashad, U. Raoof, M. Ashraf, B. Ahmed, Proportional load sharing and stability of DC microgrid with distributed architecture using SM controller. Math. Probl. Eng. **1–16**(01), 2018 (2018)
7. D. Bosich, G. Giadrossi, G. Sulligoi, S. Grillo, E. Tironi, More electric vehicles DC power systems: a large signal stability analysis in presence of CPLs fed by floating supply voltage, in *2014 IEEE International Electric Vehicle Conference (IEVC)*, pp. 1–6 (2014)
8. Y. Yoshida, H.P. Figueroa, R.A. Dougal, Comparison of energy storage configurations in railway microgrids, in *2017 IEEE Second International Conference on DC Microgrids (ICDCM)* (IEEE, New York, 2017), pp. 133–138
9. P. Magne, B. Nahid-Mobarakeh, S. Pierfederici, Active stabilization of DC microgrids without remote sensors for more electric aircraft. IEEE Trans. Ind. Appl. **49**(5), 2352–2360 (2013)
10. A. Braitor, A.R. Mills, V. Kadirkamanathan, G.C. Konstantopoulos, P.J. Norman, C.E. Jones, Control of DC power distribution system of a hybrid electric aircraft with inherent overcurrent protection, in *2018 IEEE International Conference on Electrical Systems for Aircraft, Railway, Ship Propulsion and Road Vehicles International Transportation Electrification Conference (ESARS-ITEC)* (IEEE, New York, 2018), pp. 1–6
11. P. Cairoli, R.A. Dougal, New horizons in DC shipboard power systems: new fault protection strategies are essential to the adoption of DC power systems. IEEE Electrif. Mag. **1**(2), 38–45 (2013)
12. S.K. Rao, P.J. Chauhan, S.K. Panda, G. Wilson, X. Liu, A.K. Gupta, An exercise to qualify IVAC and IVDC power system architectures for a platform supply vessel, in *2016 IEEE Transportation Electrification Conference and Expo, Asia-Pacific (ITEC Asia-Pacific)* (IEEE, New York, 2016), pp. 332–337
13. F. Zhang, C. Meng, Y. Yang, C. Sun, C. Ji, Y. Chen, W. Wei, H. Qiu, G. Yang, Advantages and challenges of DC microgrid for commercial building a case study from Xiamen university DC microgrid, in *2015 IEEE First International Conference on DC Microgrids (ICDCM)* (IEEE, New York, 2015), pp. 355–358

14. C. Jin, P. Wang, J. Xiao, Y. Tang, F.H. Choo, Implementation of hierarchical control in DC microgrids. IEEE Trans. Ind. Electron. **61**(8), 4032–4042 (2014)
15. G. Fei, K. Ren, C. Jun, Y. Tao, Primary and secondary control in DC microgrids: a review. J. Mod. Power Syst. Clean Energy **7**(2), 227–242 (2019)
16. S.E. Mhankale, A.R. Thorat, Droop control strategies of DC microgrid: a review, in *2018 International Conference on Current Trends towards Converging Technologies (ICCTCT)*, pp. 372–376 (2018)
17. C. De Persis, E.R.A. Weitenberg, F. Dörfler, A power consensus algorithm for DC microgrids. Automatica **89**, 364–375 (2018)
18. D. Marx, P. Magne, B. Nahid-Mobarakeh, S. Pierfederici, B. Davat, Large signal stability analysis tools in DC power systems with constant power loads and variable power loads-a review. IEEE Trans. Power Electron. **27**(4), 1773–1787 (2012)
19. A. Emadi, B. Fahimi, M. Ehsani, On the concept of negative impedance instability in the more electric aircraft power systems with constant power loads, in *SAE Technical Paper* (SAE International, 1999)
20. M. Su, Z. Liu, Y. Sun, H. Han, X. Hou, Stability analysis and stabilization methods of DC microgrid with multiple parallel-connected DC-DC converters loaded by CPLs. IEEE Trans. Smart Grid **9**(1), 132–142 (2018)
21. J. Zhao, F. Dörfler, Distributed control and optimization in DC microgrids. Automatica **61**, 18–26 (2015)
22. Z. Liu, S. Mei, Y. Sun, H. Han, X. Hou, J.M. Guerrero, Stability analysis of DC microgrids with constant power load under distributed control methods. Automatica **90**, 62–72 (2018)
23. S. Cuk, R. Middlebrook, Input filter considerations in design and application of switching regulators, in *Proceedings of the IEEE Industry Applications Annual Meeting* (IEEE , New York, 1976)
24. F. Duan, M. Xu, X. Yang, Y. Yao, Canonical model and design methodology for IIC DC/DC converter with constant current operation capability under shorted load. IEEE Trans. Power Electron. **31**(10), 6870–6883 (2016)
25. Z.J. Shen, Ultrafast solid-state circuit breakers: protecting converter-based AC and DC microgrids against short circuit faults [technology leaders]. IEEE Electrif. Mag. **4**(2), 70–72 (2016)
26. N. Bayati, A. Hajizadeh, M. Soltani, Protection in DC microgrids: a comparative review. IET Smart Grid **1**(3), 66–75 (2018)
27. Y. Li, X. Mao, H. Wang, C. Wen, L. Wen, An improved hiccup mode short-circuit protection technique with effective overshoot suppression for DC-DC converters. IEEE Trans. Power Electron. **28**(2), 877–885 (2013)
28. A. Tilli, C. Conficoni, Control of shunt active filters with actuation and current limits. IEEE Trans. Control Syst. Technol. **24**(2), 644–653 (2016)
29. G.C. Konstantopoulos, Q. Zhong, Current-limiting DC/DC power converters. IEEE Trans. Control Syst. Technol. **27**(2), 855–863 (2019)
30. N. Bottrell, T.C. Green, Comparison of current-limiting strategies during fault ride-through of inverters to prevent latch-up and wind-up. IEEE Trans. Power Electron. **29**(7), 3786–3797 (2014)
31. E. Coyne, D. Clarke, S. Heffernan, B. Moane, Designing ESD protection devices for ultrafast overvoltage events. IEEE Trans. Electron Devices **66**(11), 4850–4857 (2019)
32. F. Zhang, X. Duan, M. Liao, J. Zou, Z. Liu, Statistical analysis of switching overvoltages in UHV transmission lines with a controlled switching. IET Gener., Transm. Distrib. **13**(21), 4998–5004 (2019)

Chapter 2
Literature Review

In this chapter, a detailed search is performed into the existing modelling, control and theoretical stability analysis procedures of DC microgrids. The investigation starts with a brief overview of the most common converters and DC loads encountered in DC networks. A particular attention is given to current hierarchical control strategies employed in DC systems, and current methods and approaches of theoretically guaranteeing system stability. Moreover, this review looks into different ways of ensuring overvoltage and overcurrent protection within DC microgrids. The end goal is to identify potential gaps that might become future research directions.

2.1 Microgrid Components

In general, DC microgrids include a wide range of components within their structure, ranging from sources, i.e. PVs, batteries, wind turbines, electric motors etc, to different types of converters and various linear/nonlinear loads, as it can be seen in Fig. 2.1. However, in this section the focus will be kept on the elements that interface the sources to the network, namely the power converters, and the various loads that they are connected to. Particularly, the electronic circuits and mathematical models are to be presented of the most predominant converters. Furthermore, this part also includes the characteristics and mathematical models of several DC loads that one deals with most frequently in DC networks.

The methodology to obtain the dynamic models is quite straightforward and it uses, in the derivation of the power converter models, the fundamental Kirchhoff's laws. Briefly, by fixing the position of the switch (or switches), the differential equations of the circuit model are derived. Then, one combines the derived models into a single equation parametrised by the switch position function whose value must coincide, for each possible case, with the numerical values of either "zero" or "one".

A.-C. Braitor, *Advanced Hierarchical Control and Stability Analysis of DC Microgrids*, Springer Theses, https://doi.org/10.1007/978-3-030-95415-4_2

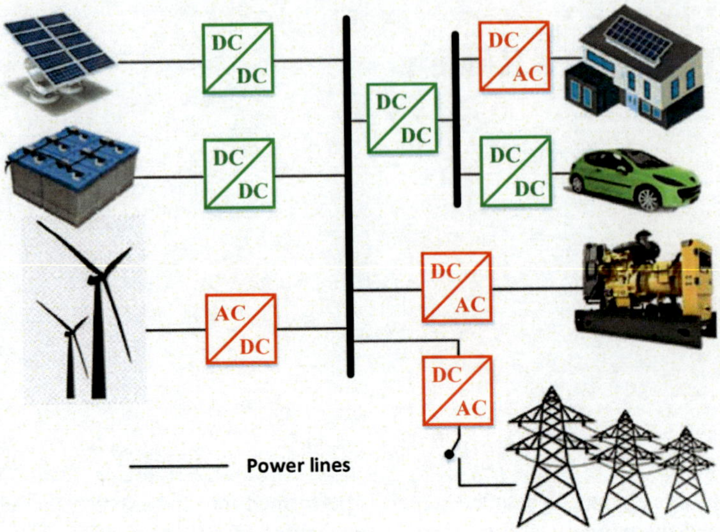

Fig. 2.1 Generic configuration of a DC microgrid

That is, the numerical values taken by the switch position function belongs to the binary set {0, 1}. The resulting model is then interpreted as an *average model* by letting the switch position function take values in the closed interval [0, 1]. This state averaging procedure has been extensively justified in the literature since the early days of power electronics. That is why, a thorough investigation into the theoretical analysis of the averaging procedure is not necessary.

Numerous textbooks in the literature present different derivations of the power converter models. A comprehensive presentation of the Euler-Lagrange modelling technique in DC/DC converters is shown in Ortega et al. [1]. Also, Kassakian et al. [2] contains detailed derivations of the most popular DC/DC power converter topologies. Standard textbooks, with a special emphasis on the steady-state PWM switched behaviour, are those of Bose [3], Mohan et al. [4] and Wood [5].

2.1.1 Power Converters

As already mentioned, the centrepiece of microgrids in general, regardless of their nature AC, DC or hybrid AC-DC, that connects the energy sources with the loads, is the power converter. In this subsection, the dynamic models are derived for several power converters present in DC microgrids. According to [6], the DC converter structures can be divided into *second order converters* (e.g. buck, boost, buck-boost, non-inverting buck-boost) and *fourth order converters* (e.g. Cuk, sepic, zeta, quadratic

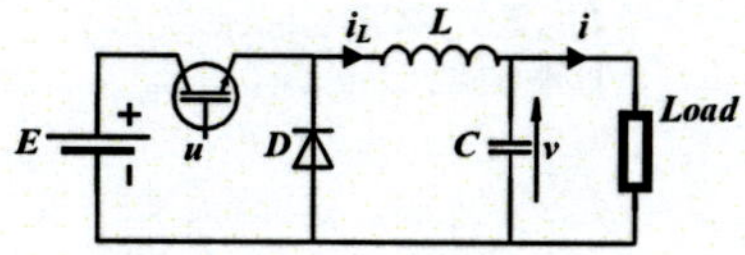

Fig. 2.2 DC/DC buck converter

buck). With respect to their number of switches, they can also be classified into two classes, i.e. *mono-variable*, or *Single Input Single Output* (SISO), and *multi-variable*, or *Multiple Input Multiple Outputs* (MIMO). One can remind the existence of converters with multiple dependent switches, which can again be classified into SISO or MIMO.

In the subsequent remainder of this section, the discussion follows around a selection of converters which are briefly introduced, and are also relevant to the work carried out in the following chapters of this thesis.

2.1.1.1 Buck Converter

Buck converters belong to the *chopper* circuits, or attenuation circuits, class. That is, the output voltage is given by the multiplication of the constant input voltage by a scalar, smaller than unity. Figure 2.2 presents the electronic circuit of the buck converter consisting of a switching element in series with the source and controlled by the control input u, bounded between $[0, 1]$, a paralleled diode D, a series inductor L and an output capacitor C, and supplying a generic load.

By applying Kirchhoff's laws in the above circuit, one obtains the average model described by the following linear differential equations:

$$L\frac{di_L}{dt} = -v + uE \tag{2.1}$$

$$C\frac{dv}{dt} = i_L - i \tag{2.2}$$

where i_L, E and i, v represent the input/output current and voltage, respectively.

The use of buck converters has been widely preferred in many theoretical microgrid applications [7–12], mainly due to their linear dynamics requiring a simpler control structure. A DC microgrid architecture is studied in [7] consisting of a DC/DC buck converter feeding a CPL, where a feedback controller combined with a feedforward strategy is proposed. In [10], the authors try to address the instability problem of DC/DC buck converters with a CPL in DC microgrids by employing a robust nonlinear controller. An offset-free model predictive controller (MPC) for DC/DC buck converters with a CPL to guarantee desired performance and ensure stability is proposed in [11]. The authors in [12] evaluate the power conversion efficiency of a buck three-level DC/DC converter, when operating in unbalanced bipolar DC microgrids.

Fig. 2.3 DC/DC boost converter

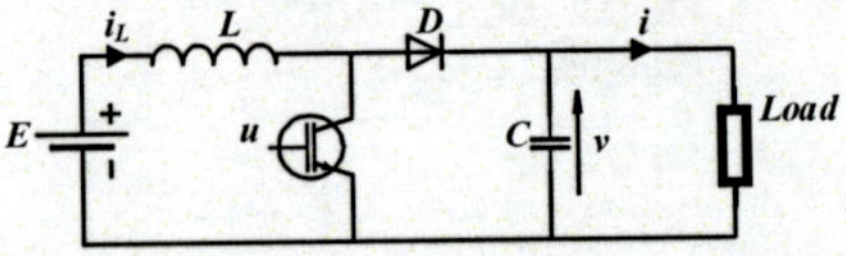

The utilisation of buck converters has been extended to various DC microgrid applications, e.g. ships [13], wind farms [14], solar farms [15], energy storage systems [16] or LVDC MG applications [17]. However, considering DC microgrids consisting exclusively of buck converters coupled to the energy sources is unrealistic, since the energy produced at the source has a low voltage and a high current and it is usually interfaced to the rest of the network by a boost converter.

2.1.1.2 Boost Converter

Since the output voltages of most distributed energy sources such as fuel cells and PV are relatively low, operating at high currents, a step-up converter is required for this sort of practical applications [18–22].

Commonly referred to as the up converter in the literature, the electronic circuit of the boost converter is depicted in Fig. 2.3. It consists of an input inductor L, with a paralleled switching element controlled by the input u, a series diode D, and an output capacitor C, supplying a generic load.

Employing Kirchhoff's laws, the boost converter dynamics are described by the following set of differential equations:

$$L\frac{di_L}{dt} = -(1-u)\,v + E \tag{2.3}$$

$$C\frac{dv}{dt} = (1-u)\,i_L - i \tag{2.4}$$

with i_L, E and i, v representing the input/output current and voltage, respectively.

Remark 1 With the similar dynamics, one can introduce the bidirectional boost converter (boost one way/buck the other way). The only difference comes from the fact that the diode D is replaced by another switching element to permit the bidirectional power-flow. The electronic circuit of the bidirectional boost is displayed in Fig. 2.4.

Fig. 2.4 Bidirectional DC/DC boost converter

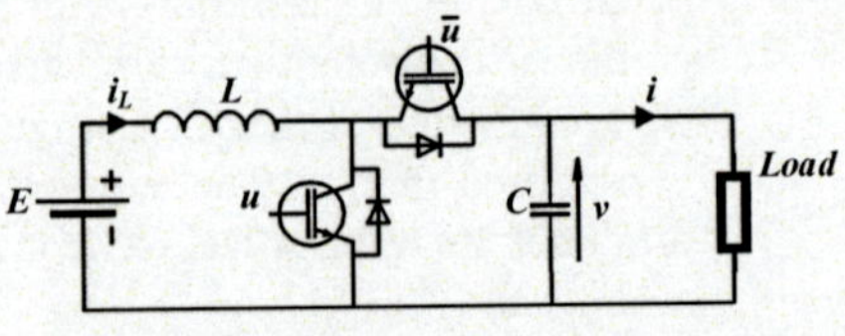

In [20], an interleaved boost converter with a magnetically coupled Cuk-type auxiliary step-up circuit that charges a voltage-doubler in the output was proposed to achieve the required voltage gain. A similar solution was presented in [21], where a sepic-integrated boost converter gives an additional step-up gain using an isolated sepic-type auxiliary step-up circuit. However, the circuit structures of the Cuk/sepic integrated high step-up converters are relatively complex and expensive, thus, they might be difficult to mass manufacture. Moreover, on the practical side, the maximum output voltage and power efficiency will be affected by the parasitic effects such as winding resistances of inductors for the Cuk/sepic integrated circuits.

The applications of both unidirectional and bidirectional converters are numerous. Droop controlled DC microgrids consisting of unidirectional boost converters have been studied in [23–25], while in [26] both unidirectional and bidirectional converters are integrated into the DC framework. A non-isolated soft-switched-integrated boost converter having high voltage gain is introduced in [27] for module-integrated PV systems, fuel cells, and other low voltage energy sources. A backstepping control strategy with finite time observers is proposed in [28], where the large signal stability of a DC microgrid having an interleaved double dual boost converter connected to a CPL is investigated. The authors in [22] present a novel modular interleaved boost converter which integrates a forward energy-delivering circuit with a voltage-doubler to achieve high step-up ratio and high efficiency for DC microgrids.

Boost converters have been successfully used in PV [29], fuel cell [30], battery [31, 32] or rural electrification [33] applications. Nevertheless, due to their non-linear dynamics, boost converter dynamics tend to be ignored in the system theoretical analysis. Most approaches use simplified or reduced-order models such as in [34, 35], others ignore the converters' dynamics completely and focus their attentions on the control framework and/or the supplied loads [36, 37].

2.1.1.3 Three-Phase Rectifier

Depending on application, as previously mentioned, PWM control techniques have been preferred to control the three-phase rectifiers in shaping the output voltage and/or current waves [38].

The schematic of the three-phase rectifier is depicted in Fig. 2.5.

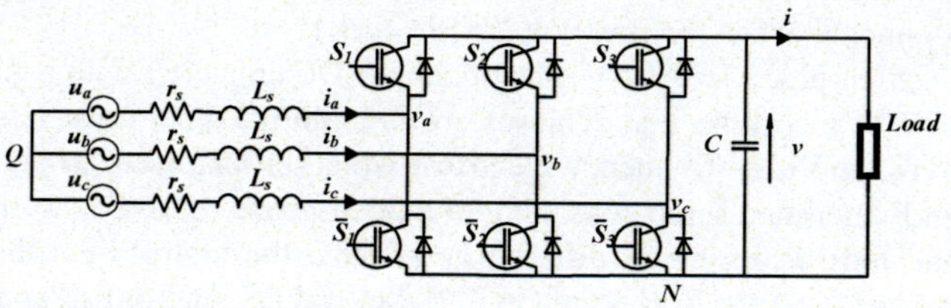

Fig. 2.5 AC/DC three-phase rectifier

The three-phase AC/DC rectifier electronic circuit has three legs with transistors, as one can notice in Fig. 2.5, and it is also known as the bidirectional boost rectifier, working with fixed DC voltage polarity.

Using Kirchhoff's laws, the dynamics of the three-phase AC/DC rectifier can be described as:

$$L_s \frac{di_a}{dt} = -r_s i_a - (v * S_1 + V_{QN}) + u_a \tag{2.5}$$

$$L_s \frac{di_b}{dt} = -r_s i_b - (v * S_2 + V_{QN}) + u_b \tag{2.6}$$

$$L_s \frac{di_c}{dt} = -r_s i_c - (v * S_3 + V_{QN}) + u_c \tag{2.7}$$

$$C \frac{dv}{dt} = (i_a * S_1) + (i_b * S_2) + (i_c * S_3) - i. \tag{2.8}$$

where the inductors L_s are selected based on their design, to achieve harmonic compensation, r_s are the parasitic resistances, while the capacitor C ensures a smooth voltage waveform; the currents i_a, i_b, i_c represent the phase currents, u_a, u_b, u_c are the three-phase sinusoidal voltages, v is the capacitor voltage, and V_{QN} represents the voltage between points N and Q. The control action S_j, with $j \in \{1, 2, 3\}$, triggers the switches, taking values from the closed interval [0, 1].

Following [39], by using Clarke and Park transformation [40], the mathematical model in synchronously rotating dq frame is set up as

$$L \frac{dI_d}{dt} = -r_s I_d - \omega L_s I_q - m_d \frac{v}{2} + U_d \tag{2.9}$$

$$L \frac{dI_q}{dt} = -r_s I_q + \omega L_s I_d - m_q \frac{v}{2} + U_q \tag{2.10}$$

$$C \frac{dv}{dt} = \frac{3}{4} m_d I_d + \frac{3}{4} m_q I_q - i. \tag{2.11}$$

with U_d, U_q and I_d, I_q being the d and q components of the grid voltages and input currents, respectively, and

$$m_d = \frac{2v_d}{v}, \quad m_q = \frac{2v_q}{v} \tag{2.12}$$

representing the duty-ratio control inputs. The voltages v_d and v_q are the d and q components of the rectifier voltage $[u_a \; u_b \; u_c]^T$.

A three-phase isolated bidirectional AC/DC converter with a modified space vector PWM algorithm that achieves AC/DC bidirectional conversion, sinusoidal AC current, and high-frequency electrical isolation has been proposed in [41]. For a slightly modified setup consisting of a three-phase AC/DC converter having $Y - \Delta$ connected transformers, a different approach in the control algorithm, but with similar improved outcome, is reported in [42]. Several DC microgrids applications incorporate three-phase rectifiers within their structure, for instance a wind energy conversion

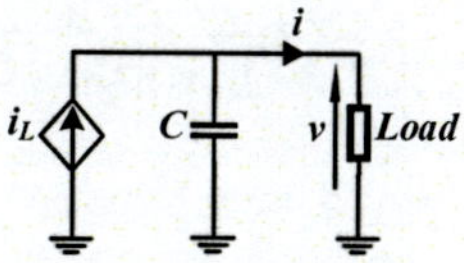

Fig. 2.6 Simplified DC network with a generic load

system (WECS) is presented in [43] where a three-phase diode bridge rectifier with a DC/DC converter has been incorporated at the terminals of a wind-driven induction generator and the DC microgrid. Microgrids with three-phase AC/DC (or DC/AC) converters have also been investigated in [44–48].

In [49, 50] the DC microgrid incorporates both bidirectional three-phase rectifiers and boost converters. However, similar to DC/DC boost converters, the nonlinear dynamic model of three-phase converters in DC microgrids with nonlinear control and nonlinear loads, following the transformation in dq frame, tends to be ignored in theoretical stability analysis, the emphasis being put on the control strategy.

2.1.2 DC Loads

In DC networks there are several types of loads one can encounter. The presence of ZIP model loads in DC microgrids, for instance, has become more and more predominant recently (see [51–57]).

Consider the electronic circuit in Fig. 2.6, consisting of a constant current source i_L, a DC capacitor C, and a generic load. In general, as explained in [58], the load I-V characteristics is given by an equation of the form:

$$i = f\,(v, c_1, c_2, \ldots, c_n) \tag{2.13}$$

where v and i represent the voltage and current drawn by the load, respectively, and $c_j = const.$, with $j \in \{1, \ldots, n\}$ are load parameters. For generic common loads, Eq. (2.13) will take different forms as shown in the following subsections.

2.1.2.1 Constant Impedance Load (CIL)

In simple terms, a constant impedance load is a load that displays an unchanging impedance, similar to a resistor. The I-V characteristics of a CIL is depicted in Fig. 2.7.

It can be observed in Fig. 2.7 that the current drawn by the CIL increases/decreases as the voltage increases/decreases. By having a constant impedance, one can notice that if the voltage changes, then the current will also change with the same ratio. A CIL can be modelled mathematically as

$$i = vG \tag{2.14}$$

where $1/G = const.$ and it represents the load impedance.

Fig. 2.7 I-V characteristics of a typical CIL

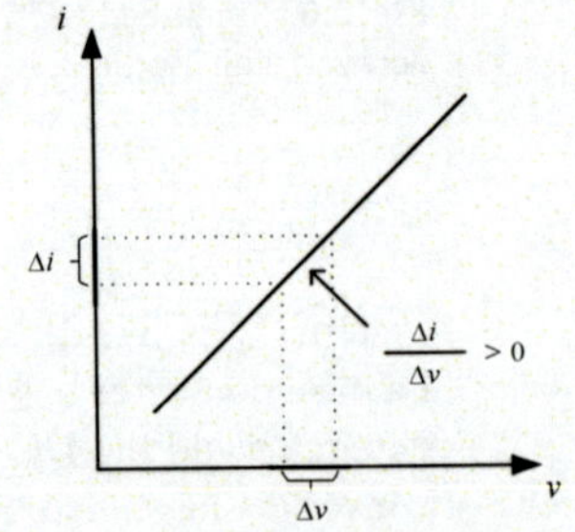

Due to its linear characteristic, it is quite straightforward to analyse its influence onto a system. A significant amount of research works, when conducting theoretical stability analysis, have utilised a CIL such as [23, 26, 37, 49, 59–62].

2.1.2.2 Constant Current Load (CCL)

A CCL draws constant current from the source.

It can be observed in Fig. 2.8 that the current drawn by the CCL remains constant as the voltage decreases/increases. A CCL can be modelled mathematically as

$$i = const. \tag{2.15}$$

One can observe that CCLs also display linear behaviour, making them easy to work with in theoretical analysis.

Remark 2 In the literature, one might also find a load much alike as the CCL, namely the constant voltage load (CVL). Since it displays a similar constant behaviour as a CCL, that is a vertical line in the I-V plane ($v = const.$), investigating it will be futile and, hence, it is only mentioned but not included in this review. Research studies with a CVL have been conducted in [63, 64].

Fig. 2.8 I-V characteristics of a typical CCL

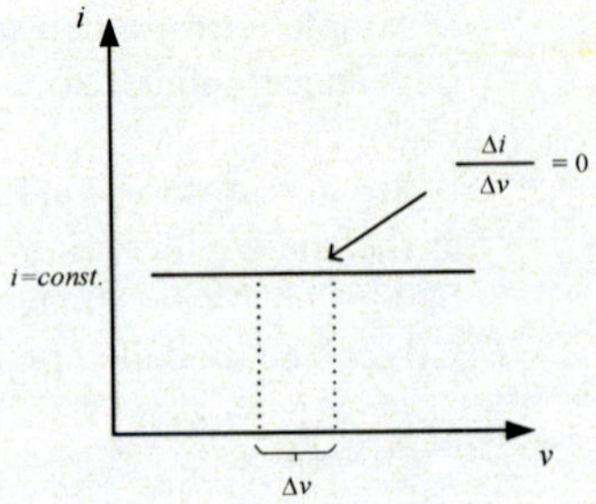

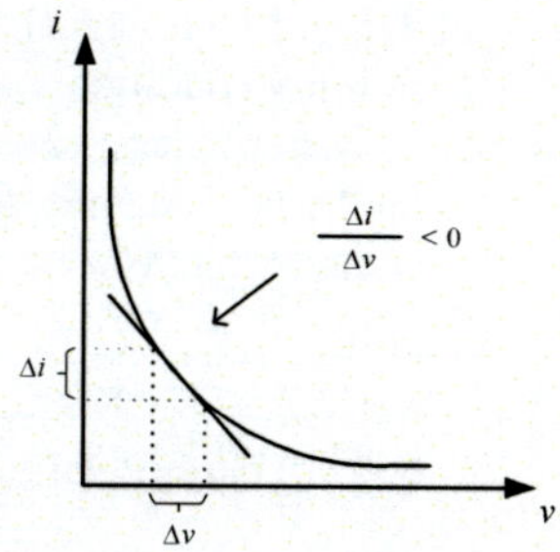

Fig. 2.9 I-V characteristics of a typical CPL

The predominance of CCLs in the literature has recorded a slight increase in the last decades. A digital implementation of a selection of loads has been done in [65], including CCL or combinations. A power-flow analysis of networks supplying loads of CCL-CPL type has been conducted in [66]. The effect of a CCL onto a DC system stability, as part of a ZIP load, has been investigated in [25]. The stability issue of ZIP loads has been also addressed in [55, 56, 67–69].

2.1.2.3 Constant Power Load (CPL)

Unlike the previously mentioned loads, CIL and CCL, a CPL exhibits negative incremental impedance, introducing instabilities in its feeder system.

It can be observed in Fig. 2.9 that the current drawn by the CPL increases/decreases as the voltage decreases/increases. A CPL can be modelled mathematically as

$$i = \frac{P}{v} \tag{2.16}$$

where $P = const.$ and it represents the power required by the load.

One can notice that CPLs have a nonlinear characteristic with a negative slope, and it behaves as a negative impedance in the small-signal analysis. That makes them complicated to analyse since it introduces additional nonlinearities into an already complex system that might include both the plant and the controller.

Remark 3 From the literature, several less common types of loads that impact the electrical system might be worth mentioning, for instance, viscous friction mechanical load [58], ventilator type mechanical load [70], etc, which are not relevant to the work done in this thesis and, hence, they have not been included.

In addressing the destabilising effect, a great deal of work has been put into research studies [9, 36, 50, 65, 71–80] of the CPLs, aiming to tackle the consequences of the negative impedance behaviour. Most approaches propose nonlinear controllers to stabilise the CPLs effect [71], or to use a circular switching surface technique [73], or to implement multistage LC filters [78] or to employ an active damping strategy [79] to satisfy the impedance inequality criteria. Moreover, small-signal [9, 36, 50,

72] and large-signal [74, 75] stability analysis are carried out to obtain sufficient stability conditions.

Nevertheless, investigating the stability of a system consisting of converters with nonlinear dynamics feeding one or more CPLs, and controlled by a nonlinear controller, remains a challenge, which has not been properly pursued so far.

2.2 Hierarchical Control

Hierarchical control strategies are fairly common in DC microgrids, and they are customarily arranged on three levels [81–83], one on top of the other, as shown in Fig. 2.10. The primary control layer deals with the voltage regulation and load sharing among units, by directly controlling the power converters. In several research works, microgrid protection, i.e. overcurrent [40, 59, 84–86], is also implemented at the primary layer. Secondary control tackles voltage fluctuations by tightening the regulation to the rated value, and improves the accuracy of the load sharing. And in most cases, secondary controllers require communication links. The tertiary control layer, or supervisory control, incorporates the energy management system. It takes care of the power flow by communicating with the lower level controllers, deciding the optimal power flow and implementing a power exchange schedule or optimal dispatch.

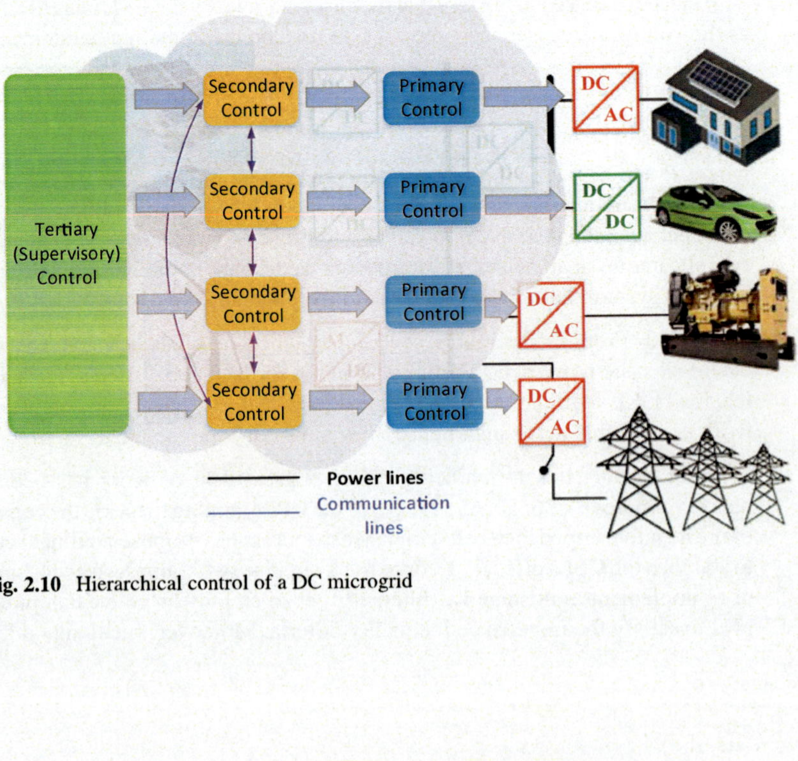

Fig. 2.10 Hierarchical control of a DC microgrid

In the following subsections, each control layer in the hierarchical control scheme will be investigated separately.

2.2.1 Primary Control

At this level, local controllers can be divided into (i) current, voltage and droop control, implemented via PI, PID or fuzzy controllers; (ii) source dependent functions such as maximum power point tracking (MPPT) applicable to PV and wind farms (for a thorough analysis of most common MPPT techniques, the reader is referred to [87, 88]) or (iii) passive and active load sharing represented by droop controllers when no communication is available, and a master slave [89] or average load sharing [90] (I-V characteristics shifted upwards) approach in the presence of communication.

From the control of power converters perspective, the lower-level primary control consists of inner loops, responsible for current and/or voltage regulation, and droop control that deals with preliminary power sharing. Regardless of the DC/DC power converter type, the inner loop control can be classified into two categories: voltage control mode and current control mode (or power control mode). The droop concept is widely applied at the primary control layer, the power being linked directly to the DC voltage. The droop characteristic of a converter in a DC microgrid can be a linear function between V and I, or similarly between V and P, soon to be explained.

A key benefit is that the conventional droop method can be easily implemented, providing voltage regulation and achieving load sharing without requiring communication links between converter units, thus, offering independent control and modularity. Its limitations, and the existing enhanced versions of the droop strategy shall be discussed in the remainder of this subsection. It is noted that most droop-based methods aim to address only one issue at a time, depending on application.

2.2.1.1 Inner Loop Control

Under voltage control mode, the DC/DC converter operates as a controllable voltage source. Similarly, in current control mode (or power control mode), the current (or power) is regulated by following a given reference, in which case the converter behaves as a controllable current (or power) source.

The voltage loop in voltage control mode is displayed in Fig. 2.11a. The voltage V is regulated to the reference V^*, and the output of the voltage controller G_V is the duty cycle u. In Fig. 2.11b, an inner current control loop is depicted when the converter operates in current control mode. In this case, the duty cycle u is acquired through a current regulator G_I. A cascaded voltage and current loop is shown in Fig. 2.11c. Starting with a set voltage reference V^*, the voltage controller G_V provides the inductor current reference i^*. The control signal u is provided by the current regulator G_I.

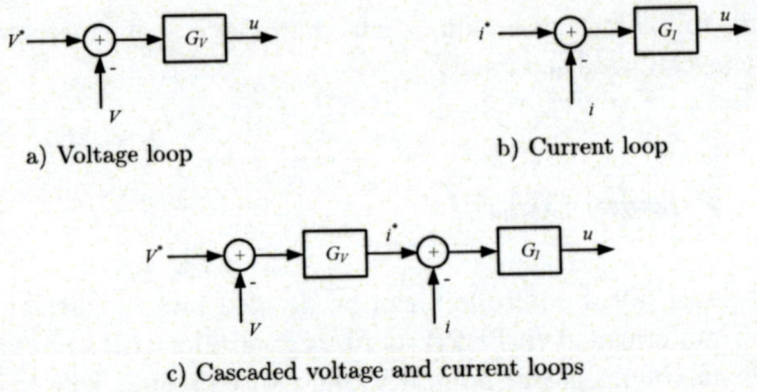

Fig. 2.11 Inner loop control of DC/DC converters

The design and analysis of inner loops are fundamental for adequate operation of the islanded microgrids (see [91–93]). Cascaded inner control approach for DC microgrids has been reported in [94], and also used for comparison with the proposed controllers in [95].

2.2.1.2 Droop Control

As one can tell from the graph in Fig. 2.12, the droop expression is given by the static equation:

$$v = V^* - \Delta v \tag{2.17}$$

where v represents the output voltage set to be regulated close to the reference V^*, while the output voltage deviation is $\Delta v = mi$, with m being the droop coefficient, while i representing the output current.

It is clearly noted that the conventional droop strategy presents some serious disadvantages stemmed from the voltage regulation and power sharing trade-off. Hence, to improve the conventional droop form, numerous droop control variations have been developed to cope with the shortcomings that the conventional droop exhibits, and eventually enhance the overall system performance.

Fig. 2.12 Conventional droop control V-I characteristic

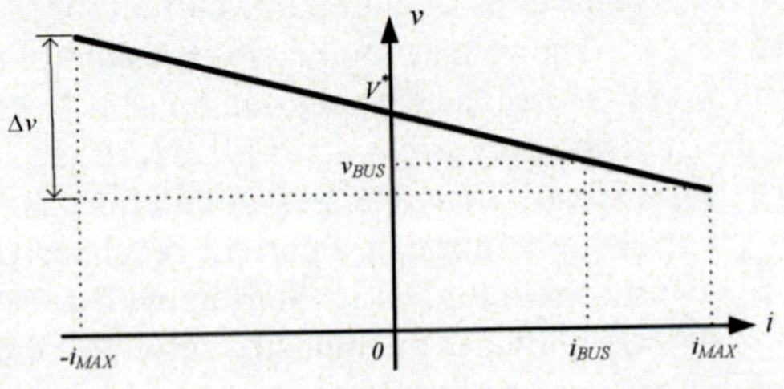

The problem of optimal, nonlinear and distributed designs of droop controls in DC microgrids has been thoroughly studied in [96]. The authors conclude that the resulting control is optimal only under the assumption that full information of the microgrid is available. By relaxing this requirement leads to a reduction in the control optimality, one of which portrays precisely the conventional droop control.

Here are some variations of the conventional droop methods:

1. Inverse droop
   First presented in [97], the decentralised inverse-droop control achieves power sharing for input-series-output-parallel (ISOP) DC/DC converters. The advantage of the proposed inverse-droop is that the output voltage rises as the load power demand increases. The difference from the conventional one is that the feedback polarity of the output current is positive instead of negative.
   More recent research works (see [98, 99]), that have employed the inverse-droop control for ISOP DC/DC converters, also aimed to investigate small-signal stability using Routh-Hurwitz criterion and plotting root locus diagrams. Although numerical approaches are important, the vital valuable insights would come from an analytical proof which, for now, continues to remain a challenge However, as one can tell from the current literature, the application of the inverse droop strategy, so far, is limited to the particular ISOP type DC/DC converters. This fact eliminates its potential universal uses.
2. Robust droop
   The robust droop controller used for a DC distribution network, as explained in [100], uses the load terminal voltage as a feedback signal. Furthermore, the line impedance is treated as part of the equivalent output impedance of individual power converters, while the inaccuracy of load sharing is minimised by regulating the newly introduced robust coefficients. A similar approach is presented in [101] for a hybrid interlinking converter of an AC/DC microgrid.
   According to the graph in Fig. 2.13, the robust droop expression becomes

$$v_o = V^* - \Delta v_o, \tag{2.18}$$

   where v_o represents the load voltage set to be regulated close to the reference V^*, with the load voltage deviation being $\Delta v_o = mi$.
   A robust droop controller of grid-connected inverters, aimed to cancel the effects of changes in the grid side impedance values, based on disturbance observer is

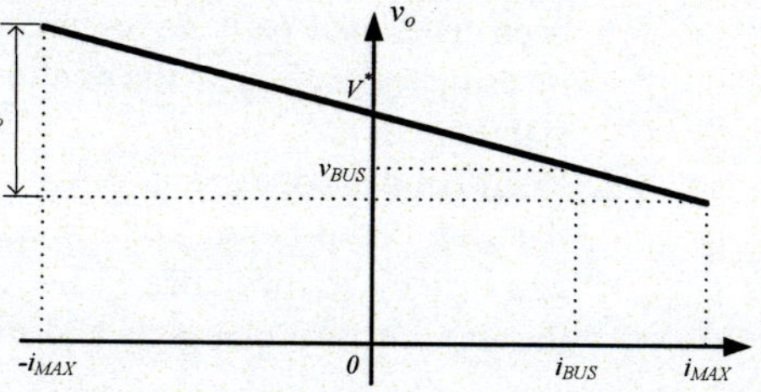

Fig. 2.13 Robust droop control V-I characteristic

proposed in [102]. In [103], a parallel configuration inverter network is studied, each unit equipped with a robust droop multiple loop controller, that includes an outer power loop and inner voltage and current loops. The outer loop adopts the robust droop control strategy to improve the accuracy of the power sharing, while the inner loops employ a quasi-proportional resonant (QPR) controller to track the output voltage, and restrain the circulating current.

However, the main disadvantage of the robust droop control strategy is that it requires the measurement of the terminal bus voltage, which may not always be accessible, and also the measurement of the output currents, which needs additional sensors. Also, as shown in [100], in order for the robust droop strategy to be successfully implemented, the robust coefficient has to be chosen within a certain range to avoid instabilities in the system.

3. Quadratic droop

   In [104, 105], a droop-like voltage feedback controller is introduced that is quadratic in the local voltage magnitude, permitting in the closed-loop system the use of circuit-theory analysis techniques. One can claim that the quadratic droop controller is just a special case of the general feedback controller, but a key fact is that the authors manage to show a one-to-one correspondence between the high-voltage equilibrium under the proposed method, and the solution of an optimisation problem that minimises the trade-off between reactive power dissipation and voltage deviations.

   The stability of voltage dynamics for a power network, in which nodal voltages are controlled by means of quadratic droop controllers with nonlinear AC reactive power as inputs, is investigated in [106]. But again, the stringent lack of an analytical proof of stability including the entire closed-loop system remains an unsolved problem, and a potential path to pursue.

4. Nonlinear droop

   An interesting approach in [107] employs a nonlinear droop control strategy to control the charge/discharge of a supercapacitor whenever required by the grid. An improved nonlinear droop is introduced in [108], which also includes a comparison of different droop control strategies. The authors in [109] design a nonlinear output feedback linearised droop control to guarantee active and reactive power sharing in a parallel configuration network. The procedure of constructing the nonlinear droops to minimise the operating costs and share the output power effectively among the sources has been proposed in [110].

   On one side, the effect that the nonlinear droop expressions incorporate into the stability of the microgrids has never been studied. In [111] a numerical approach to ensure system stability by testing automatically different operating point has been developed. Still, no research work has looked into the matter of analytically investigate the effect of the nonlinear droop expression on the closed-loop system stability.

5. Dead-band droop

   An useful drop-based strategy is the dead-band droop, proposed for energy storage systems to introduce a "floating" or standby working mode, thus, avoiding the unnecessary charging and discharging cycles. A droop-controlled micro-

grid that integrates energy storage systems with dead-band droop controllers is studied in [112, 113].
But, the use for the standby operating mode is restricted to energy storage systems. Thus, the applications of dead-band droop controllers are limited, whilst their implementation is deemed inadequate for most converters within the DC microgrid framework.
6. Adaptive droop
Adaptive droop controllers [24, 111, 112, 114–117] are fairly common at the primary control layer of DC microgrids. An adaptive droop control strategy is introduced in [118], where a variable droop coefficient strategy is able to adjust the ratio of energy storage unit power online, reduce the DC bus voltage fluctuations and improve the system stability. A master-slave control strategy is proposed in [119] based on the adaptive droop controller of the voltage source converter. The adaptive droop control in [120], forces the units to share the power according to the available headroom of each converter station. An interesting adaptive integrated coordinated control is presented in [121], consisting of an adaptive droop control and an internal current loop.
However, yet again the discussion of the analytical proof of stability for the entire closed-loop system intervenes, since similar to previous droop control methods, it is not properly addressed, nor studied.

To improve the overall performance of the droop-based strategies, secondary controllers are often incorporated on top of the primary control layer. A more detailed view is given in the following subsection.

2.2.2 Secondary Control

As pointed out previously, droop controllers implemented at the primary control layer impose a performance trade-off between power sharing accuracy and voltage regulation. Numerous centralised, decentralised and distributed approaches in hierarchical control frameworks have emerged in the last years to boost the reliability of DC microgrids and compensate the voltage deviation and improve the power sharing accuracy [36, 81, 82, 115, 122–132]. By exchanging information via communication links, the secondary control is able to adjust the set-points of the primary control. This approach is not new as it has been utilised in the early network-based control of power electronic systems [133–135]. The effect of time-delays was investigated in [135], where a radio frequency communication-based control for interactive power electronics networks was introduced. One might agree that due to the distributed nature of microgrids, low bandwidth communication (LBC) is best suitable for the implementation of secondary controllers as it can be put into place using the controllers area network, power line communication, and others [136].

Whilst, as already mentioned, in most cases primary controller are implemented in a decentralised topology, upper layer controllers on the other hand emerged in sev-

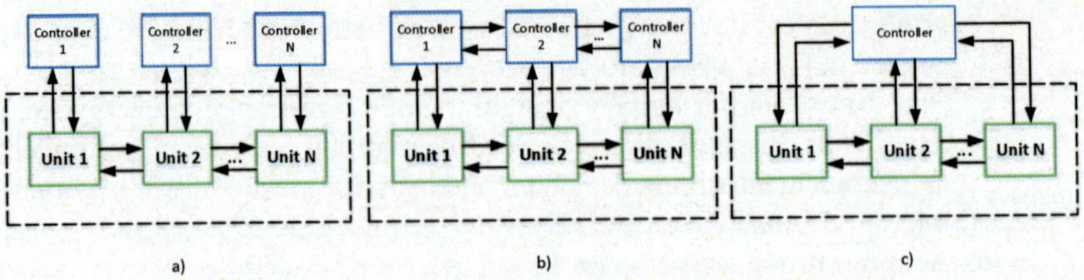

Fig. 2.14 Control architectures for multiple unit networks: **a** decentralised; **b** distributed; **c** centralised

eral architectures depending on application. Most common control architectures of multiple unit networks are depicted in Fig. 2.14, including centralised, decentralised and distributed approaches.

2.2.2.1 Centralised Control

A conventional way of realizing power sharing between batteries and supercapacitors is through measuring the total unbalanced power using a centralised controller and splitting it into high and low frequency components [137, 138]. As highlighted in [139], centralised control is effective only if there is no communication delay, provided energy storage units operate in power control mode. To tackle the effect of the communication delays, supercapacitors are expected to regulate the bus voltage, while the battery would compensate for the low-frequency component of the unbalanced power, as shown in [140].

A centralised secondary control method is proposed in [81], where the secondary control adjustment is completed in the microgrid central controller, which restores the voltage to the rated value using low-bandwidth communication network. The load voltage is compared to the reference value, and the difference δV is sent through the LBC to all modules to compensate the voltage difference. The droop expression becomes:

$$v = V^* + \delta V - Ri. \tag{2.19}$$

with δV representing the voltage difference $V^* - v$. A similar approach is presented in [114] to avoid voltage deviations, where a virtual resistance is changed inside the centralised controller to transit criteria for changing unit-level operating modes. The centralised controller in [114] adjusts the droop gains for the bidirectional DC/DC converters that interface the batteries such that they keep identical state of charge in islanded mode. And during charging, a higher droop coefficient is allocated to the battery with a higher state of charge, whereas when discharging, a highest droop coefficient is assigned to the battery with the lowest state of charge. Centralised controllers have also been proposed in [141–144].

However, the centralised control framework relies on the central controller and it is easily affected if subjected to a failure. This vulnerability makes this approach unsuitable as it is not robust enough for future DC microgrid applications.

2.2.2.2 Decentralised Control

To overcome the drawbacks that the centralised controller introduces, decentralised methods have been applied [116, 125, 145–148]. When employing this control architecture, the units are autonomously controlled using real-time feedbacks. Decentralised approaches are implemented without the need of a communication network, hence, are achieved exclusively by local controllers. A common technique is the DC bus signalling, where the information carrier is the DC bus [149]. Although this method is communication-free and fairly straightforward, it is also inaccurate due to the error in the DC bus voltage estimation. Furthermore, since they lack the data from other converter units, these controllers have an inherent performance limitation.

Another decentralised control technique is the power line signalling that injects sinusoidal signal of specific frequency at the DC bus, as shown in [150]. A decentralised control strategy named frequency coordinating virtual impedance control is proposed in [139], and it introduces a low-pass filter (LPF) or high-pass filter (HPF) to the droop control in order to coordinate power sources with different characteristics. Nevertheless, it causes a voltage deviation not addressed in these control frameworks. In [147], a high-pass-filter-based droop control is integrated in battery controllers in order to compensate for the voltage deviation. But, this strategy is not suitable for microgrids with multiple hybrid energy storage systems.

A unified decentralised control strategy has been proposed in [151] that can provide an energy management and ensure the voltage restoration to the rated value. In grid-connected mode, the converters' output current is controlled through an outer current loop with a PI controller, which is also useful for the economic dispatch in the microgrid. In islanded mode, the reference value for the output currents is modified by an average voltage controller based on PI.

2.2.2.3 Distributed Control

The authors in [82] present a distributed secondary control approach, where the secondary control scheme is implemented in the local controller, while the data used in the local secondary control framework is exchanged via an LBC network. Compared to the centralised approach, a key benefit is that this strategy avoids the scenario of single-point-of-failure. However, the effect of line resistance has not been comprehensively considered. Furthermore, to guarantee an accurate power sharing one has to choose large droop coefficients which affects the voltage regulation.

A different secondary control scheme is presented in [124], where two additional control methods are proposed to average the output voltage and output current. These control loops are integrated locally, guaranteeing those two tasks simultaneously. The main disadvantage is that every converter requires information from all the other converters. This strategy can be regarded as static averaging, which is not flexible and leads to high communication stress. In [115, 129], a consensus algorithm with a sparse communication network is proposed, where only the information from the neighbouring converters is required. In doing so, the flexibility of the system is enhanced, giving a plug-and-play feature, and since the global information from all converters is no longer required, the communication stress is highly reduced.

A novel distributed control strategy is introduced in [113] where the authors consider the utilisation priority of the energy storage, renewable and non-renewable sources. The control approach in [152] utilises a distributed control algorithm and the concept of multiple slack buses, that enables the microgrid's redundancy, modularity and expandability, while also ensuring the plug-and-play feature. Numerous distributed approaches for DC microgrids have emerged in the past decade, such as multi-agent based control for operation cost minimisation [153], consensus algorithms to facilitate the modelling and sensitivity analysis [154], power management based on fuzzy logic [155], or power management based on DC bus signalling [156].

2.2.3 Supervisory Control

A triple-role supervisory control strategy was introduced in [114], on top of existing primary control for a DC microgrid consisting of renewable sources and batteries. Its first feature includes an online adaptation of the virtual resistances designed to achieve asymptotic approaching of batteries' state-of-charge (SOC) and is intended for moderate replenishment periods. At high SOCs, the second and third feature kick in, being responsible for distributing the charging/discharging tokens and transitions of operating modes, respectively.

An optimal planning of DC sources by properly sharing the load among the generator units and synchronizing individual incremental costs is presented in [157]. In [158], the introduced method optimally adjusts the set-points of the DG units, batteries, RES and interlinking converter to restore the voltage and frequency to nominal values. In doing so, the generation cost is reduced at short time intervals. The power management issue in a DC microgrid has been tackled in [159] where a supervisory control scheme has been proposed to improve the bus voltage deviation and power sharing; similar results in [160] by using a fuzzy logic based supervisory control, in [161] via a multiagent supervisory control approach, or in [162] by employing an algorithm that accounts for seamless operation under various modes and state of charge limit conditions of hybrid energy storage systems.

Economic optimal dispatch and scheduling implemented through an upper layer supervisory control has also attracted a lot of interest in past decade. In [163] a multi-time scale rolling optimal dispatch with day-ahead scheduling has been proposed. The robustness given by the day-ahead scheduling is also encountered in [164] where the proposed algorithm converts the min-max-min problem of each level into a two-stage mixed-integer linear programming problem, and similarly in [165]. An optimal power dispatch strategy for cost and emission reduction under load and generation uncertainties is introduced in [166]. Supervisory controllers to reach a particular optima have also been proposed in [167–170].

2.3 Microgrid Stability

To cope with the instability issues that CPL exhibit, several studies [34–37, 171] have been carried out not only in terms of modelling and control, but with a particular emphasis on the stability theory part. Most approaches can be divided into three categories depending on the number of converters taken into consideration when designing the DC microgrid, that is single converter, two converter or more converters.

Stability analysis and stabilisation strategies of one converter feeding a CPL that fall into the first category include [8, 75, 77, 78, 172–177]. One notices that by increasing the damping and reducing the negative impedance, the destabilizing effect of the CPL, namely the voltage oscillations, can be mitigated [77, 173]. The interconnection and damping assignment technique have been used for stability analysis of a DC microgrid [77]. The describing function method to analyse stability is reported in [178]. The authors in [75] propose several tools for large-signal stability which include Brayton and Moser's, nonlinear circuit theory [78, 172, 174], block diagonalised Lyapunov function, semi-tensor product of matrices [179] and reverse trajectory tracking. Other solutions to solve the stability problem are sliding-mode control [8], Popov criterion [175], nonlinear feedback [176], phase-plane analysis [177].

An interesting strategy aimed towards a multi-converter DC microgrid to obtain the stability region by using an LMI approach is proposed in [9, 36, 180]. The use of behavioural modeling technique is applied in the stability analysis as showcased in [181].

2.3.1 Stability of Droop-Controlled DC Microgrids

The second category involves cases when one has two converters in parallel [72, 182–187]. Amplitude death solutions are utilised for stabilisation of DC microgrids with instantaneous CPLs in [182]. Eigenvalue analysis is employed in [13]. The nonlinear feedback method is applied here as well for two parallel converters [183]. The so-called MIMO generalised Bode criterion is proposed in [188]. The stability analysis of two parallel converters with V-I droop control is carried out in [189].

Synergetic control has been introduced in [184], and it is based on several properties of nonlinear dynamic dissipative systems. A comparison between [183, 184] has been done in [185]. Linear methods based on virtual impedance are proposed in [72, 186, 187] to increase damping. However, unlike [72], the proposed strategies in [186, 187] have also incorporated a voltage restoration term. And by contrast, the nonlinear methods [182–184] have a wider stability region compared to the linear methods [72, 186, 187], but it requires global information, such as the size of the CPLs.

The third group extends the complexity of the DC microgrid to n number of converters [34, 35, 190–192]. The stability of a reduced-order model has been studied in [34, 35], obtaining a stable domain for the droop coefficients. A region-based stability analysis is proposed in [190] to achieve a reasonable selection of parameters. An extended stability analysis method based on Floquet theory is presented in [191]. Similarly, in [192], the authors investigate system stability by making use of the monodromy matrix, whose eigenvalues are the Floquet multipliers. Considering a master/slave configuration in [193], it is revealed that the interactions among the master converter and the slave converters will cause instability under certain conditions.

A detailed review on the stability studies carried out for paralleled DC/DC converters is showcased in [194]. Nevertheless, in the majority of these cases, the dynamics of the converters are not included in the analysis, and only a reduced-order model or the control system alone is analysed.

2.3.2 *Stability of DC Microgrids under Secondary Control*

Stability analysis of microgrid clusters under primary and secondary control has been presented in [55]. The stability of DC microgrid clusters has also been investigated in [128]. In [36, 37], the stability analysis of DC microgrids under distributed secondary control has been studied, acquiring sufficient stability conditions for the controller parameters. But in these works, only the secondary controller dynamics have been considered, the microgrid system and primary controller dynamics being ignored. The stability of a decentralised communication-free secondary voltage restoration and current sharing control for islanded DC microgrids is studied in [195]. Likewise, a hybrid secondary controller is introduced in [36] to achieve both current sharing and voltage regulation.

The authors in [171] derive two stability criteria under different conditions for the DC system, considering two communication time delays, namely constant delay and time-varying delay, respectively. An extensive review on stabilisation techniques in hierarchical controlled DC microgrids has been put together in [196].

2.4 Microgrid Protection

The emergence of DERs is setting the stage for modern power systems to operate as MGs, avoiding power disruptions and permitting a fast recovery during grid disturbances. Therefore, MGs are vital assets to have in order to improve grid resilience, which is seriously undermined if MGs are not properly protected in the event of faults [197].

In [198], an analysis for different relay types has been conducted by considering different faults and generation conditions in a MG. A protection scheme based on information sharing technology is proposed in [199], the protection devices having the ability to communicate with each other and share their local information. The effect of replacing undervoltage protection with a differential protection scheme is studied in [200].

2.4.1 Overcurrent Protection

A critical issue, tightly related to the stability analysis, and corresponding to the technical limitations of the system components is the current-limiting property of the converter units. *Overcurrent* (or excess current) instances occur when an electric current larger than intended passes through a conductor producing excessive generation of heat, presenting a high risk of fire or damage to equipment. As described in [85], a current limitation would protect the components without violating fixed limits, set by the technical requirements of each converter. An adaptive overcurrent protection for distribution networks has been developed in [201, 202]. A neural network based overcurrent protection approach is shown in [203].

Thus, apart from ensuring system stability, the equipment's overcurrent protection must be guaranteed at all times, by complying to its technical limitations. Such matter is crucial particularly during transients, or operating under faulty conditions or unrealistic power demands. Even though the converter overcurrent protection is customarily taken care of by utilising extra fuses, circuit breakers and protection relays, a recorded increase has been noticed in focus and interest to design control strategies able to guarantee an inherent current-limiting property [204]. Existing current limitation methods can adjust the original control framework to ensure overcurrent protection [205], but closed-loop stability cannot be analytically proven. Moreover, the original control method might be prone to integrator wind-up and latch-up cases which again can yield instability [205].

2.4.2 Overvoltage Protection

Overvoltage instances occur when the voltage in a circuit, or part of the circuit, increases above its designed limit, causing potential damages and faults in the converter components or the grid.

The methods proposed in [206, 207] aim to reduce the active power injected by a source until voltage complies with the operation requirements, a strategy known as active power curtailment (APC). Potential limitations and main challenges one faces, when applying this method, have been discussed in [208], and they include:

- decreasing active power injection in one phase may lead to a decrease in voltage in the injection phase, but it will increase the voltage in the other phases;
- using the APC ON/OFF active power control approach to avoid the occurrence of overvoltages, the sequence of solving a problem but creating a new one leads to system instability;
- APC strategy may cause the sequential loss of the interconnected DER, triggered by the first shut-down (a harmful event, commonly referred to as *cascading*).

In [209], a comparison using optimal power flow between centralised and local voltage control solutions have been conducted to mitigate the voltage rise impact. The authors in [210] introduce a methodology to identify and locate transient overvoltages using wavelet packet decomposition (WPD) and general regression neural networks (GRNN) theory.

But, since all these strategies are implemented at the upper control layers, which operate in a higher time-scale, they are usually slower than the primary control layer. So, managing to implement an overvoltage protection at the primary control layer might be more efficient and an important step forward.

2.5 Gaps in the Current Literature

Network topology is key in microgrids. Currently in the literature, the computation of the admittance matrix in DC microgrids, as shown in [9, 36], does not consider the instantaneous nonlinear expressions of the output currents and the load voltage. Instead, it uses the linearised power balance equation to obtain the load voltage with respect to its steady-state equilibrium. This voltage is then replaced into the output currents expressions, which are further used to compute the admittance matrix as a function of the equilibrium point of the load voltage.

Moving to the control part, the disadvantages of conventional droop control strategy are well known, and its improved versions exhibit the following shortcomings:

- several droop-based variations are designed for specific applications, i.e. inversed droop strategy is limited to the use of ISOP DC/DC converters only, while dead-band droop controllers are only applicable to energy storage systems;
- the robust droop controller is based on several assumptions, such as the availability of the load voltage measurement. It also requires extra sensors to measure the output currents and it introduces a gain restricted in a limited domain that may affect system stability;
- other droop methods, i.e. quadratic, nonlinear and adaptive droop, lack a complete theoretical proof of stability, which is of paramount importance in the theoretical analysis of DC microgrids.

With the primary control posing so many challenges, a multi-level control scheme is often required. In most cases, hierarchical control schemes that implement centralised upper layer controllers are not disturbance robust as they suffer from single point of failure, and are vulnerable to faults. Decentralised architectures could be a better solution, despite their local implementation, whereas distributed control approaches although more efficient, they require communication.

When it comes to investigating the overall stability of the closed-loop system, most approaches rely on strong assumptions that lead to limitations:

- most studies use extensively buck DC/DC converters which have linear dynamics;
- several approaches are based on reduced-order models [34, 35];
- in some cases, the converters' dynamics are completely ignored, and only the controller dynamics are investigated [36, 37];
- others employ numerical approaches, such as the root locus.

Unlike conventional methods that ensure the protection of microgrids through different pieces of hardware, there is an increased interest in implementing it through control. At the moment:

- in microgrid applications, conventional overcurrent protection devices are being utilised, such as fuses and relays;
- in some cases the current limitation is guaranteed by using saturation units, which are ineffective during transients;
- so far, overvoltage protection is implemented only at the upper layers of the hierarchical control architecture.

References

1. R. Ortega, A. Loria, P.J. Nicklasson, H. Sira-Ramirez, *Passivity-Based Control of Euler-Lagrange Systems: Mechanical, Electrical and Electromechanical Applications* (Springer, Berlin, 1998)
2. G. Verghese, J. Kassakian, M. Schlecht, *Principles of Power Electronics* (Addison-Wesley, Reading, 1991)
3. B. Bose, *Modern Power Electronics and AC Drives* (Prentice Hall, Englewood Cliffs, 2001)
4. N. Mohan, T.M. Undeland, W.P. Robbins, *Power Electronic Converter, Applications and Design*, 3rd edn. (Wiley, New York, 2003)
5. P. Wood, *Switching Power Converters* (Van Nostrand Reinhold Co., New York, 1981)
6. H.J. Sira Ramirez, R. Silva-Ortigoza, *Control Design Techniques in Power Electronics Devices* (Springer, London, 2006)
7. J. Solsona, S. GomezJorge, C. Busada, Nonlinear control of a buck converter feeding a constant power load. IEEE Lat. Am. Trans. **12**(5), 899–903 (2014)
8. Y. Zhao, W. Qiao, D. Ha, A sliding-mode duty-ratio controller for DC/DC buck converters with constant power loads. IEEE Trans. Ind. Appl. **50**(2), 1448–1458 (2014)
9. M. Su, Z. Liu, Y. Sun, H. Han, X. Hou, Stability analysis and stabilization methods of DC microgrid with multiple parallel-connected DC-DC converters loaded by CPLs. IEEE Trans. Smart Grid **9**(1), 132–142 (2018)
10. M.A. Hassan, E. Li, X. Li, T. Li, C. Duan, S. Chi, Adaptive passivity-based control of DC-DC buck power converter with constant power load in DC microgrid systems. IEEE J. Emerg. Sel. Top. Power Electron. **7**(3), 2029–2040 (2019)

11. Q. Xu, Y. Yan, C. Zhang, T. Dragicevic, F. Blaabjerg, An offset-free composite model predictive control strategy for DC/DC buck converter feeding constant power loads. IEEE Trans. Power Electron. **35**(5), 5331–5342 (2020)
12. G.V. Broeck, W. Martinez, M. DallaVecchia, S. Ravyts, J. Driesen, Conversion efficiency of the buck three-level DC-DC converter in unbalanced bipolar DC microgrids. IEEE Trans. Power Electron. **35**(9), 9306–9319 (2020)
13. Y. Ma, K. Corzine, A. Maqsood, F. Gao, K. Wang, Stability assessment of droop controlled parallel buck converters in zonal ship DC microgrid, in *2019 IEEE Electric Ship Technologies Symposium (ESTS)* (IEEE, New York, 2019), pp. 268–272
14. S.S. Gite, S.H. Pawar, Modeling of wind energy system with MPPT control for DC microgrid, in *2017 Second International Conference on Electrical, Computer and Communication Technologies (ICECCT)*, pp. 1–6 (2017)
15. B.R. Jogi, B.Y. K, A.A. Kadle, A DC-DC buck converter for standalone solar DC microgrid system using perturb and observe MPPT algorithm, in *2019 Global Conference for Advancement in Technology (GCAT)*, pp. 1–5 (2019)
16. S.A. Kumar, M.N. Suneetha, C. Lakshminarayana, The simulation study of solar PV coupled synchronous buck converter with lead acid battery charging, in *2019 International Conference on Power Electronics Applications and Technology in Present Energy Scenario (PETPES)*, pp. 1–4 (2019)
17. N. Rigogiannis, D. Baros, E. Lempesis, S. Ntoga, E. Peioglou, C. Pechlivanis, N. Papanikolaou, Experimental investigation of a digitally current controlled synchronous buck DC/DC converter for microgrids applications, in *2019 Panhellenic Conference on Electronics Telecommunications (PACET)*, pp. 1–5 (2019)
18. S. Choe, J. Ahn, J. Lee, S. Baek, Dynamic simulator for a PEM fuel cell system with a PWM DC/DC converter. IEEE Trans. Energy Convers. **23**(2), 669–680 (2008)
19. J. Lai, Power conditioning circuit topologies. IEEE Ind. Electron. Mag. **3**(2), 24–34 (2009)
20. C. Pan, C. Lai, A high-efficiency high step-up converter with low switch voltage stress for fuel-cell system applications. IEEE Trans. Ind. Electron. **57**(6), 1998–2006 (2010)
21. K. Park, G. Moon, M. Youn, Nonisolated high step-up boost converter integrated with sepic converter. IEEE Trans. Power Electron. **25**(9), 2266–2275 (2010)
22. C. Lai, C. Pan, M. Cheng, High-efficiency modular high step-up interleaved boost converter for DC-microgrid applications. IEEE Trans. Ind. Appl. **48**(1), 161–171 (2012)
23. A.C. Braitor, G.C. Konstantopoulos, V. Kadirkamanathan, Power sharing of parallel operated DC-DC converters using current-limiting droop control, in *2017 25th Mediterranean Conference on Control and Automation (MED)*, pp. 528–533 (2017)
24. M. Mokhtar, M.I. Marei, A.A. El-Sattar, An adaptive droop control scheme for DC microgrids integrating sliding mode voltage and current controlled boost converters. IEEE Trans. Smart Grid **10**(2), 1685–1693 (2019)
25. A.C. Braitor, G.C. Konstantopoulos, V. Kadirkamanathan, Current-limiting droop control design and stability analysis for paralleled boost converters in DC microgrids. IEEE Trans. Control. Syst. Technol. **29**(1), 385–394 (2020)
26. A. Braitor, A.R. Mills, V. Kadirkamanathan, G.C. Konstantopoulos, P.J. Norman, C.E. Jones, Control of DC power distribution system of a hybrid electric aircraft with inherent overcurrent protection, in *2018 IEEE International Conference on Electrical Systems for Aircraft, Railway, Ship Propulsion and Road Vehicles International Transportation Electrification Conference (ESARS-ITEC)* (IEEE, New York, 2018), pp. 1–6
27. S. Sathyan, H.M. Suryawanshi, B. Singh, C. Chakraborty, V. Verma, M.S. Ballal, ZVS-ZCS high voltage gain integrated boost converter for DC microgrid. IEEE Trans. Ind. Electron. **63**(11), 6898–6908 (2016)
28. Q. Xu, W. Jiang, F. Blaabjerg, C. Zhang, X. Zhang, T. Fernando, Backstepping control for large signal stability of high boost ratio interleaved converter interfaced DC microgrids with constant power loads. IEEE Trans. Power Electron. **35**(5), 5397–5407 (2020)
29. M. Kumar, B. Tyagi, Design of a model reference adaptive controller (MRAC) for DC-DC boost converter for variations in solar outputs using modified MIT rule in an islanded microgrid, in *2020 IEEE International Conference on Power Electronics, Smart Grid and Renewable Energy (PESGRE2020)* (IEEE, New York, 2020), pp. 1–6

30. M. Alam, K. Kumar, J. Srivastava, V. Dutta, A study on DC microgrids voltages based on photovoltaic and fuel cell power generators, in *2018 7th International Conference on Renewable Energy Research and Applications (ICRERA)*, pp. 643–648 (2018)
31. P.S. Tomar, A.K. Sharma, K. Hada, Energy storage in DC microgrid system using non-isolated bidirectional soft-switching DC/DC converter, in *2017 6th International Conference on Computer Applications in Electrical Engineering-Recent Advances (CERA)*, pp. 439–444 (2017)
32. L. Saeed, M.Y. Ali Khan, S.H. Khan, M. Azhar, Design and control for a multiport DC-DC boost converter with battery backup for microgrid, in *2019 International Conference on Engineering and Emerging Technologies (ICEET)*, pp. 1–6 (2019)
33. S. Saha, A. Bhattacharjee, D. Elangovan, G. Arunkumar, DC microgrid system for rural electrification, in *2017 International Conference on Energy, Communication, Data Analytics and Soft Computing (ICECDS)*, pp. 307–313 (2017)
34. S. Anand, B.G. Fernandes, Reduced-order model and stability analysis of low-voltage DC microgrid. IEEE Trans. Ind. Electron. **60**(11), 5040–5049 (2013)
35. A.P.N. Tahim, D.J. Pagano, E. Lenz, V. Stramosk, Modeling and stability analysis of islanded DC microgrids under droop control. IEEE Trans. Power Electron. **30**(8), 4597–4607 (2015)
36. Z. Liu, S. Mei, Y. Sun, H. Han, X. Hou, J.M. Guerrero, Stability analysis of DC microgrids with constant power load under distributed control methods. Automatica **90**, 62–72 (2018)
37. F. Guo, Q. Xu, C. Wen, L. Wang, P. Wang, Distributed secondary control for power allocation and voltage restoration in islanded DC microgrids. IEEE Trans. Sustain. Energy **9**(4), 1857–1869 (2018)
38. N.A. Rahim, T.C. Green, B.W. Williams, PWM ASIC design for the three-phase bi-directional buck converter. Int. J. Electron. **81**(5), 603–615 (1996)
39. T.-S. Lee, Lagrangian modeling and passivity-based control of three-phase AC/DC voltage-source converters. IEEE Trans. Ind. Electron. **51**(4), 892–902 (2004)
40. Q.C. Zhong, G.C. Konstantopoulos, Current-limiting three-phase rectifiers. IEEE Trans. Ind. Electron. **65**(2), 957–967 (2018)
41. L. Gu, K. Jin, A three-phase isolated bidirectional AC/DC converter and its modified SVPWM algorithm. IEEE Trans. Power Electron. **30**(10), 5458–5468 (2015)
42. L. Gu, K. Jin, A three-phase bidirectional AC/DC converter with star-delta connected transformers. IEEE Trans. Power Electron. **31**(12), 8115–8125 (2016)
43. V. Nayanar, N. Kumaresan, N. AmmasaiGounden, A single-sensor-based MPPT controller for wind-driven induction generators supplying DC microgrid. IEEE Trans. Power Electron. **31**(2), 1161–1172 (2016)
44. C.G.C. Branco, R.P. Torrico-Bascope, C.M.T. Cruz, F.K.A. de Lima, Proposal of three-phase high-frequency transformer isolation UPS topologies for distributed generation applications. IEEE Trans. Ind. Electron. **60**(4), 1520–1531 (2013)
45. X. Guo, Z. Lu, B. Wang, X. Sun, L. Wang, J.M. Guerrero, Dynamic phasors-based modeling and stability analysis of droop-controlled inverters for microgrid applications. IEEE Trans. Smart Grid **5**(6), 2980–2987 (2014)
46. J.A. Mueller, M. Rasheduzzaman, J.W. Kimball, A model modification process for grid-connected inverters used in islanded microgrids. IEEE Trans. Energy Convers. **31**(1), 240–250 (2016)
47. J. Schiffer, T. Seel, J. Raisch, T. Sezi, Voltage stability and reactive power sharing in inverter-based microgrids with consensus-based distributed voltage control. IEEE Trans. Control Syst. Technol. **24**(1), 96–109 (2016)
48. J. Choi, A. Khalsa, D.A. Klapp, S. Baktiono, M.S. Illindala, Survivability of prime-mover powered inverter-based distributed energy resources during microgrid islanding. IEEE Trans. Ind. Appl. **55**(2), 1214–1224 (2019)
49. A. Braitor, P.R. Baldivieso-Monasterios, G.C. Konstantopoulos, V. Kadirkamanathan, Current-limiting droop control design of paralleled AC/DC and DC/DC converters in DC micro-grids, in *IECON 2018—44th Annual Conference of the IEEE Industrial Electronics Society* (IEEE, New York, 2018), pp. 132–137

50. A. Braitor, G.C. Konstantopoulos, V. Kadirkamanathan, Stability analysis and nonlinear current-limiting control design for DC micro-grids with CPLs. IET Smart Grid **3**(3), 355–366 (2020)
51. Y. Ge, A.J. Flueck, D. Kim, J. Ahn, J. Lee, D. Kwon, An event-oriented method for online load modeling based on synchrophasor data. IEEE Trans. Smart Grid **6**(4), 2060–2068 (2015)
52. M. Gheydi, A. Nouri, N. Ghadimi, Planning in microgrids with conservation of voltage reduction. IEEE Syst. J. **12**(3), 2782–2790 (2018)
53. L. Liang, Y. Hou, D.J. Hill, Enhancing flexibility of an islanded microgrid with electric springs. IEEE Trans. Smart Grid **10**(1), 899–909 (2019)
54. Z. Shuai, Y. Peng, X. Liu, Z. Li, J.M. Guerrero, Z.J. Shen, Parameter stability region analysis of islanded microgrid based on bifurcation theory. IEEE Trans. Smart Grid **10**(6), 6580–6591 (2019)
55. R. Han, M. Tucci, A. Martinelli, J.M. Guerrero, G. Ferrari-Trecate, Stability analysis of primary plug-and-play and secondary leader-based controllers for DC microgrid clusters. IEEE Trans. Power Syst. **34**(3), 1780–1800 (2019)
56. B. Fan, S. Guo, J. Peng, Q. Yang, W. Liu, L. Liu, A consensus-based algorithm for power sharing and voltage regulation in DC microgrids. IEEE Trans. Ind. Inform. **16**(6), 3987–3996 (2020)
57. S. Taheri, V. Kekatos, Power flow solvers for direct current networks. IEEE Trans. Smart Grid **11**(1), 634–643 (2020)
58. K.Y. Khouzam, Optimum load matching in direct-coupled photovoltaic power systems-application to resistive loads. IEEE Trans. Energy Convers. **5**(2), 265–271 (1990)
59. G.C. Konstantopoulos, Q. Zhong, Current-limiting DC/DC power converters. IEEE Trans. Control Syst. Technol. **27**(2), 855–863 (2019)
60. N. Zhi, H. Zhang, N. Li, J. Yang, System-level design and stability analysis of DC microgrid, in *2014 International Power Electronics and Application Conference and Exposition*, pp. 1134–1137 (2014)
61. N. Ghanbari, S. Bhattacharya, M. Mobarrez, Modeling and stability analysis of a DC microgrid employing distributed control algorithm, in *2018 9th IEEE International Symposium on Power Electronics for Distributed Generation Systems (PEDG)*, pp. 1–7 (2018)
62. X. Liu, Y. Bian, Large signal stability analysis of the DC microgrid with the storage system, in *2017 20th International Conference on Electrical Machines and Systems (ICEMS)*, pp. 1–5 (2017)
63. X. Qu, H. Han, S. Wong, C.K. Tse, W. Chen, Hybrid IPT topologies with constant current or constant voltage output for battery charging applications. IEEE Trans. Power Electron. **30**(11), 6329–6337 (2015)
64. S. Arora, P. Balsara, D. Bhatia, Input-output linearization of a boost converter with mixed load (constant voltage load and constant power load). IEEE Trans. Power Electron. **34**(1), 815–825 (2019)
65. S. Arora, P.T. Balsara, D.K. Bhatia, Digital implementation of constant power load (CPL), active resistive load, constant current load and combinations, in *2016 IEEE Dallas Circuits and Systems Conference (DCAS)* (IEEE, New York, 2016), pp. 1–4
66. M. Hoger, M. Siranec, J. Altus, Power flow analysis of networks with loads of constant current - constant power factor type, in *2019 20th International Scientific Conference on Electric Power Engineering (EPE)*, pp. 1–4 (2019)
67. M. Cucuzzella, R. Lazzari, Y. Kawano, K.C. Kosaraju, J.M.A. Scherpen, Robust passivity-based control of boost converters in DC microgrids, in *2019 IEEE 58th Conference on Decision and Control (CDC)* (IEEE, New York, 2019), pp. 8435–8440
68. M. Cucuzzella, K.C. Kosaraju, J.M.A, Scherpen, Passivity-based voltage control of DC microgrids: addressing the stability issue of zip loads, in *2020 European Control Conference (ECC)*, pp. 298–301 (2020)
69. P. Nahata, G. Ferrari-Trecate, On existence of equilibria, voltage balancing, and current sharing in consensus-based DC microgrids, in *2020 European Control Conference (ECC)*, pp. 1216–1223 (2020)

70. M.M. Saied, Matching of DC motors to photovoltaic generators for maximum daily gross mechanical energy. IEEE Trans. Energy Convers. **3**(3), 465–472 (1988)
71. Q. Xu, C. Zhang, C. Wen, P. Wang, A novel composite nonlinear controller for stabilization of constant power load in DC microgrid. IEEE Trans. Smart Grid **10**(1), 752–761 (2019)
72. X. Lu, K. Sun, J.M. Guerrero, J.C. Vasquez, L. Huang, J. Wang, Stability enhancement based on virtual impedance for DC microgrids with constant power loads. IEEE Trans. Smart Grid **6**(6), 2770–2783 (2015)
73. M. Anun, M. Ordonez, I.G. Zurbriggen, G.G. Oggier, Circular switching surface technique: high-performance constant power load stabilization for electric vehicle systems. IEEE Trans. Power Electron. **30**(8), 4560–4572 (2015)
74. D. Bosich, G. Giadrossi, G. Sulligoi, S. Grillo, E. Tironi, More electric vehicles DC power systems: a large signal stability analysis in presence of CPLs fed by floating supply voltage, in *2014 IEEE International Electric Vehicle Conference (IEVC)* (IEEE, New York, 2014), pp. 1–6
75. D. Marx, P. Magne, B. Nahid-Mobarakeh, S. Pierfederici, B. Davat, Large signal stability analysis tools in DC power systems with constant power loads and variable power loads-a review. IEEE Trans. Power Electron. **27**(4), 1773–1787 (2012)
76. G. Sulligoi, D. Bosich, L. Zhu, M. Cupelli, A. Monti, Linearizing control of shipboard multi-machine MVDC power systems feeding constant power loads, in *2012 IEEE Energy Conversion Congress and Exposition (ECCE)* (IEEE, New York, 2012), pp. 691–697
77. A. Kwasinski, C.N. Onwuchekwa, Dynamic behavior and stabilization of DC microgrids with instantaneous constant-power loads. IEEE Trans. Power Electron. **26**(3), 822–834 (2011)
78. X. Liu, Y. Zhou, W. Zhang, S. Ma, Stability criteria for constant power loads with multistage *lc* filters. IEEE Trans. Veh. Technol. **60**(5), 2042–2049 (2011)
79. A.M. Rahimi, A. Emadi, Active damping in DC/DC power electronic converters: a novel method to overcome the problems of constant power loads. IEEE Trans. Ind. Electron. **56**(5), 1428–1439 (2009)
80. A. Emadi, A. Khaligh, C.H. Rivetta, G.A. Williamson, Constant power loads and negative impedance instability in automotive systems: definition, modeling, stability, and control of power electronic converters and motor drives. IEEE Trans. Veh. Technol. **55**(4), 1112–1125 (2006)
81. J.M. Guerrero, J.C. Vasquez, J. Matas, L.G. de Vicuna, M. Castilla, Hierarchical control of droop-controlled AC and DC microgrids 2014; a general approach toward standardization. IEEE Trans. Ind. Electron. **58**(1), 158–172 (2011)
82. S. Anand, B.G. Fernandes, J. Guerrero, Distributed control to ensure proportional load sharing and improve voltage regulation in low-voltage DC microgrids. IEEE Trans. Power Electron. **28**(4), 1900–1913 (2013)
83. C.N. Papadimitriou, E.I. Zountouridou, N.D. Hatziargyriou, Review of hierarchical control in DC microgrids. Electr. Power Syst. Res. **122**, 159–167 (2015)
84. Q.C. Zhong, G.C. Konstantopoulos, Current-limiting droop control of grid-connected inverters. IEEE Trans. Ind. Electron. (99), 1–1 (2016)
85. G.C. Konstantopoulos, Q.C. Zhong, Nonlinear control of DC/DC power converters with inherent current and power limitation, in *2016 24th Mediterranean Conference on Control and Automation (MED)*, pp. 949–954 (2016)
86. G.C. Konstantopoulos, Q.C. Zhong, W.L. Ming, PLL-less nonlinear current-limiting controller for single-phase grid-tied inverters: design, stability analysis, and operation under grid faults. IEEE Trans. Ind. Electron. **63**(9), 5582–5591 (2016)
87. B. Subudhi, R. Pradhan, A comparative study on maximum power point tracking techniques for photovoltaic power systems. IEEE Trans. Sustain. Energy **4**(1), 89–98 (2013)
88. M.A.G. de Brito, L. Galotto, L.P. Sampaio, G.d.A.e Melo, C.A. Canesin, Evaluation of the main MPPT techniques for photovoltaic applications. IEEE Trans. Ind. Electron. **60**(3), 1156–1167 (2013)
89. Y. Panov, J. Rajagopalan, F.C. Lee, Analysis and design of N paralleled DC-DC converters with master-slave current-sharing control, in *Proceedings of APEC 97 - Applied Power Electronics Conference*, vol. 1, pp. 436–442 (1997)

90. W. Qiu, Z. Liang, Practical design considerations of current sharing control for parallel VRM applications, in *Twentieth Annual IEEE Applied Power Electronics Conference and Exposition, 2005. APEC 2005*, vol. 1, pp. 281–286 (2005)
91. F. de Bosio, L.A. De S. Ribeiro, M.S. Lima, F. Freijedo, J.M. Guerrero, M. Pastorelli, Inner current loop analysis and design based on resonant regulators for isolated microgrids, in *2015 IEEE 13th Brazilian Power Electronics Conference and 1st Southern Power Electronics Conference (COBEP/SPEC)*, pp. 1–6 (2015)
92. G.W. Ndiwulu, E. De Jaeger, A.K. Lusala, Inner control loops approach to control the islanded photovoltaic microgrid, in *2017 IEEE AFRICON*, pp. 1161–1166 (2017)
93. Y. Peng, T. Yin, M. Li, Y. Wang, D. Hu, Z. Liu, A sequence impedance modeling of VSG with consideration of inner loops control, in *2019 4th IEEE Workshop on the Electronic Grid (eGRID)* (IEEE, New York, 2019), pp. 1–5
94. A. Elnady, PI controller based operational scheme to stabilize voltage in microgrid, in *2019 Advances in Science and Engineering Technology International Conferences (ASET)*, pp. 1–6 (2019)
95. H.U. Rahman Habib, S. Wang, M.F. Elmorshedy, A. Waqar, R.M. Imran, K.M. Kotb, Performance enhancement of power converters for PV-based microgrid using model predictive control, in *2019 International Conference on Electrical, Communication, and Computer Engineering (ICECCE)*, pp. 1–6 (2019)
96. A. Maknouninejad, Z. Qu, F.L. Lewis, A. Davoudi, Optimal, nonlinear, and distributed designs of droop controls for DC microgrids. IEEE Trans. Smart Grid **5**(5), 2508–2516 (2014)
97. G. Xu, D. Sha, X. Liao, Decentralized inverse-droop control for input-series-output-parallel DC-DC converters. IEEE Trans. Power Electron. **30**(9), 4621–4625 (2015)
98. L. Shu, W. Chen, X. Jiang, Decentralized control for fully modular input-series output-parallel (ISOP) inverter system based on the active power inverse-droop method. IEEE Trans. Power Electron. **33**(9), 7521–7530 (2018)
99. S.H. Kim, B.J. Kim, C.Y. Won, A study on decentralized inverse-droop control for input voltage sharing of ISOP converter in the current control loop, in *2019 10th International Conference on Power Electronics and ECCE Asia (ICPE 2019 - ECCE Asia)*, pp. 2382–2387 (2019)
100. Z. Shuai, D. He, J. Fang, Z.J. Shen, C. Tu, J. Wang, Robust droop control of DC distribution networks. IET Renew. Power Gener. **10**(6), 807–814 (2016)
101. F. Alam, M. Ashfaq, S.S. Zaidi, A.Y. Memon, Robust droop control design for a hybrid AC/DC microgrid, in *2016 UKACC 11th International Conference on Control (CONTROL)*, pp. 1–6 (2016)
102. M. Elkayam, A. Kuperman, J.M. Guerrero, Robust droop control of grid-connected inverters, in *2016 IEEE International Conference on the Science of Electrical Engineering (ICSEE)*, pp. 1–4 (2016)
103. Y. Chen, A. Luo, Z. Peng, Robust droop multiple loop control method for resistance output parallel inverters, in *2nd IET Renewable Power Generation Conference (RPG 2013)*, pp. 1–4 (2013)
104. J.W. Simpson-Porco, F. Dörfler, F. Bullo, Voltage stabilization in microgrids via quadratic droop control. IEEE Trans. Autom. Control **62**(3), 1239–1253 (2017)
105. J.W. Simpson-Porco, F. Dörfler, F. Bullo, Voltage stabilization in microgrids via quadratic droop control, in *52nd IEEE Conference on Decision and Control*, pp. 7582–7589 (2013)
106. M. Jafarian, H. Sandberg, K.H. Johansson, The interconnection of quadratic droop voltage controllers is a Lotka-Volterra system: implications for stability analysis. IEEE Control Syst. Lett. **2**(2), 218–223 (2018)
107. D. Hulea, O. Cornea, N. Muntean, Nonlinear droop charging control of a supercapacitor with a bi-directional hybrid DC-DC converter, in *2016 IEEE 16th International Conference on Environment and Electrical Engineering (EEEIC)*, pp. 1–6 (2016)
108. H. Liang, Y. Huang, Z. Liu, Research on nonlinear drooping control strategy of DC microgrid, in *TENCON 2018 - 2018 IEEE Region 10 Conference*, pp. 0865–0870 (2018)

109. R. Mahmud, M.A. Hossain, H. Pota, Nonlinear output feedback droop control for parallel inverters in standalone microgrids, in *2019 9th International Conference on Power and Energy Systems (ICPES)*, pp. 1–6 (2019)
110. A. Elrayyah, Y. Sozer, Construction of nonlinear droop relations to optimize islanded microgrids operation, in *2013 IEEE Energy Conversion Congress and Exposition*, pp. 1663–1668 (2013)
111. H. Abdelgabir, A.R. Boynuegri, A. Elrayyah, Y. Sozer, A complete small signal modelling and adaptive stability analysis of nonlinear droop-controlled microgrids, in *2018 IEEE Applied Power Electronics Conference and Exposition (APEC)*, pp. 3333–3339 (2018)
112. J.-W. Kim, H.-S. Choi, B.H. Cho, A novel droop method for converter parallel operation. IEEE Trans. Power Electron. **17**(1), 25–32 (2002)
113. C. Jin, P. Wang, J. Xiao, Y. Tang, F.H. Choo, Implementation of hierarchical control in DC microgrids. IEEE Trans. Ind. Electron. **61**(8), 4032–4042 (2014)
114. T. Dragičević, J.M. Guerrero, J.C. Vasquez, D. Škrlec, Supervisory control of an adaptive-droop regulated DC microgrid with battery management capability. IEEE Trans. Power Electron. **29**(2), 695–706 (2014)
115. V. Nasirian, A. Davoudi, F.L. Lewis, J.M. Guerrero, Distributed adaptive droop control for DC distribution systems. IEEE Trans. Energy Convers. **29**(4), 944–956 (2014)
116. X. Lu, K. Sun, J.M. Guerrero, J.C. Vasquez, L. Huang, State-of-charge balance using adaptive droop control for distributed energy storage systems in DC microgrid applications. IEEE Trans. Ind. Electron. **61**(6), 2804–2815 (2014)
117. Y.A.I. Mohamed, E.F. El-Saadany, Adaptive decentralized droop controller to preserve power sharing stability of paralleled inverters in distributed generation microgrids. IEEE Trans. Power Electron. **23**(6), 2806–2816 (2008)
118. H. Yu, X. Feng, H. Liu, Z. Lv, C. Zhou. Research on an adaptive droop control used in DC micro-grid, in *2016 IEEE 8th International Power Electronics and Motion Control Conference (IPEMC-ECCE Asia)*, pp. 2634–2638 (2016)
119. X. Huo, X. Xie, Z. Wu, X. Cao, A novel DC voltage control strategy for DC distribution based on adaptive droop control, in *2020 4th International Conference on Green Energy and Applications (ICGEA)*, pp. 40–44 (2020)
120. Y. Wang, H. Jiang, L. Zhou, P. Xing, An improved adaptive droop control strategy for power sharing in micro-grid, in *2016 8th International Conference on Intelligent Human-Machine Systems and Cybernetics (IHMSC)*, vol. 01, pp. 50–53 (2016)
121. Z. Huang, Adaptive integrated coordinated control strategy for MMC-MTDC systems, in *2018 International Conference on Power System Technology (POWERCON)*, pp. 2440–2447 (2018)
122. D. Salomonsson, L. Soder, A. Sannino, An adaptive control system for a DC microgrid for data centers. IEEE Trans. Ind. Appl. **44**(6), 1910–1917 (2008)
123. D. Chen, L. Xu, L. Yao, DC voltage variation based autonomous control of DC microgrids. IEEE Trans. Power Deliv. **28**(2), 637–648 (2013)
124. X. Lu, J.M. Guerrero, K. Sun, J.C. Vasquez, An improved droop control method for DC microgrids based on low bandwidth communication with DC bus voltage restoration and enhanced current sharing accuracy. IEEE Trans. Power Electron. **29**(4), 1800–1812 (2014)
125. A. Khorsandi, M. Ashourloo, H. Mokhtari, A decentralized control method for a low-voltage DC microgrid. IEEE Trans. Energy Convers. **29**(4), 793–801 (2014)
126. Q. Shafiee, J.M. Guerrero, J.C. Vasquez, Distributed secondary control for islanded microgrids-a novel approach. IEEE Trans. Power Electron. **29**(2), 1018–1031 (2014)
127. L. Che, M. Shahidehpour, DC microgrids: economic operation and enhancement of resilience by hierarchical control. IEEE Trans. Smart Grid **5**(5), 2517–2526 (2014)
128. Q. Shafiee, T. Dragičević, J.C. Vasquez, J.M. Guerrero, Hierarchical control for multiple DC-microgrids clusters. IEEE Trans. Energy Convers. **29**(4), 922–933 (2014)
129. V. Nasirian, S. Moayedi, A. Davoudi, F.L. Lewis, Distributed cooperative control of DC microgrids. IEEE Trans. Power Electron. **30**(4), 2288–2303 (2015)

130. S. Li, Q. Yu, H. Zhang, S. Gao, Y. Song, D. Wang, J. Tang, A hierarchical multi-agent evaluation scheme for integrating distributed energy resources in distribution systems, in *International Conference on Renewable Power Generation (RPG 2015)*, pp. 1–6 (2015)
131. H. Han, Y. Liu, Y. Sun, M. Su, J.M. Guerrero, An improved droop control strategy for reactive power sharing in islanded microgrid. IEEE Trans. Power Electron. **30**(6), 3133–3141 (2015)
132. S. Ahmadi, D. Nazarpour, Q. Shafiee, H. Bevrani, A fuzzy inference model for distributed secondary control of islanded microgrids, in *2016 24th Iranian Conference on Electrical Engineering (ICEE)*, pp. 1233–1238 (2016)
133. K. Acharya, M. Tahir, S.K. Mazumder, Communication fault-tolerant wireless network control of a load-sharing multiphase interactive power network, in *2006 37th IEEE Power Electronics Specialists Conference*, pp. 1–8 (2006)
134. S.K. Mazumder, M. Tahir, K. Acharya, Master-slave current-sharing control of a parallel DC-DC converter system over an RF communication interface. IEEE Trans. Ind. Electron. **55**(1), 59–66 (2008)
135. S.K. Mazumder, K. Acharya, M. Tahir, Joint optimization of control performance and network resource utilization in homogeneous power networks. IEEE Trans. Ind. Electron. **56**(5), 1736–1745 (2009)
136. P. Wang, X. Lu, X. Yang, W. Wang, D. Xu, An improved distributed secondary control method for DC microgrids with enhanced dynamic current sharing performance. IEEE Trans. Power Electron. **31**(9), 6658–6673 (2016)
137. M. Farhadi, O. Mohammed, Adaptive energy management in redundant hybrid DC microgrid for pulse load mitigation. IEEE Trans. Smart Grid **6**(1), 54–62 (2015)
138. B. Liu, F. Zhuo, Y. Zhu, H. Yi, System operation and energy management of a renewable energy-based DC micro-grid for high penetration depth application. IEEE Trans. Smart Grid **6**(3), 1147–1155 (2015)
139. Y. Gu, W. Li, X. He, Frequency-coordinating virtual impedance for autonomous power management of DC microgrid. IEEE Trans. Power Electron. **30**(4), 2328–2337 (2015)
140. D. Chen, L. Xu, Autonomous DC voltage control of a DC microgrid with multiple slack terminals. IEEE Trans. Power Syst. **27**(4), 1897–1905 (2012)
141. M. Gulin, M. VaŠak, T. Pavlović, Dynamical behaviour analysis of a DC microgrid in distributed and centralized voltage control configurations, in *2014 IEEE 23rd International Symposium on Industrial Electronics (ISIE)*, pp. 2365–2370 (2014)
142. A.J. Babqi, Z. Yi, A.H. Etemadi, Centralized finite control set model predictive control for multiple distributed generator small-scale microgrids, in *2017 North American Power Symposium (NAPS)*, pp. 1–5 (2017)
143. M. Mehdi, C. Kim, M. Saad, Robust centralized control for DC islanded microgrid considering communication network delay. IEEE Access **8**, 77765–77778 (2020)
144. M. Saleh, Y. Esa, A. Mohamed, Centralized control for DC microgrid using finite state machine, in *2017 IEEE Power Energy Society Innovative Smart Grid Technologies Conference (ISGT)*, pp. 1–5 (2017)
145. Q. Xu, J. Xiao, X. Hu, P. Wang, M.Y. Lee, A decentralized power management strategy for hybrid energy storage system with autonomous bus voltage restoration and state-of-charge recovery. IEEE Trans. Ind. Electron. **64**(9), 7098–7108 (2017)
146. Q. Xu, X. Hu, P. Wang, J. Xiao, P. Tu, C. Wen, M.Y. Lee, A decentralized dynamic power sharing strategy for hybrid energy storage system in autonomous DC microgrid. IEEE Trans. Ind. Electron. **64**(7), 5930–5941 (2017)
147. Q. Xu, J. Xiao, P. Wang, X. Pan, C. Wen, A decentralized control strategy for autonomous transient power sharing and state-of-charge recovery in hybrid energy storage systems. IEEE Trans. Sustain. Energy **8**(4), 1443–1452 (2017)
148. P. Lin, P. Wang, J. Xiao, J. Wang, C. Jin, Y. Tang, An integral droop for transient power allocation and output impedance shaping of hybrid energy storage system in DC microgrid. IEEE Trans. Power Electron. **33**(7), 6262–6277 (2018)
149. F. Li, Z. Lin, Z. Qian, J. Wu, Active DC bus signaling control method for coordinating multiple energy storage devices in DC microgrid, in *2017 IEEE Second International Conference on DC Microgrids (ICDCM)*, pp. 221–226 (2017)

150. IEEE standard for power-line carrier line-tuning equipment (30 khz to 500 khz) associated with power transmission lines. IEEE Std C93.4-2012, pp. 1–67 (2013)
151. Y. Xia, Y. Peng, H. Hu, Y. Wang, W. Wei, Advanced unified decentralised control method with voltage restoration for DC microgrids. IET Renew. Power Gener. **10**(6), 861–871 (2016)
152. M. Mobarrez, N. Ghanbari, S. Bhattacharya, Control hardware-in-the-loop demonstration of a building-scale dc microgrid utilizing distributed control algorithm, in *2018 IEEE Power Energy Society General Meeting (PESGM)*, pp. 1–5 (2018)
153. C. Li, J.C. Vasquez, J.M. Guerrero, Multiagent-based distributed control for operation cost minimization of droop controlled DC microgrid using incremental cost consensus, in *IECON 2015 - 41st Annual Conference of the IEEE Industrial Electronics Society*, pp. 005202–005205 (2015)
154. L. Meng, T. Dragicevic, J.C. Vasquez, J.M. Guerrero, J.R. Pérez, Modeling and sensitivity analysis of consensus algorithm based distributed hierarchical control for DC microgrids, in *2015 IEEE Applied Power Electronics Conference and Exposition (APEC)*, pp. 342–349 (2015)
155. R.R. Deshmukh, M.S. Ballal, G.G. Talapur, H.M. Suryawanshi, Distributed control for power management based on fuzzy logic in DC microgrid, in *2018 2nd IEEE International Conference on Power Electronics, Intelligent Control and Energy Systems (ICPEICES)*, pp. 1–6 (2018)
156. D. Jeong, H. Yun, H. Kim, H. Kim, J. Baek, Distributed control strategy of DC microgrid for islanding mode operation, in *2017 19th European Conference on Power Electronics and Applications (EPE'17 ECCE Europe)*, pp. P.1–P.5 (2017)
157. V. Nasirian, F.L. Lewis, A. Davoudi, Distributed optimal dispatch for DC distribution networks, in *2015 IEEE First International Conference on DC Microgrids (ICDCM)*, pp. 97–101 (2015)
158. M. Khalid, Discussion on -novel supervisory control method for islanded droop-based AC/DC microgrids-. IEEE Trans. Power Syst. **35**(5), 4138–4138 (2020)
159. R.R. Deshmukh, M.S. Ballal, A supervisory modified control scheme for power management in multi bus DC microgrid, vol. 1, pp. 2282–2287 (2019)
160. R.R. Deshmukh, M.S. Ballal, H. Suryawanshi, A fuzzy logic based supervisory control for power management in multibus DC microgrid. IEEE Trans. Ind. Appl. 1–1 (2020)
161. A.A. Hamad, M.A. Azzouz, E.F. El-Saadany, Multiagent supervisory control for power management in DC microgrids. IEEE Trans. Smart Grid **7**(2), 1057–1068 (2016)
162. R.K. Sharma, S. Mishra, Dynamic power management and control of a PV PEM fuel-cell-based standalone AC/DC microgrid using hybrid energy storage. IEEE Trans. Ind. Appl. **54**(1), 526–538 (2018)
163. H. Qiu, W. Gu, Y. Xu, B. Zhao, Multi-time-scale rolling optimal dispatch for AC/DC hybrid microgrids with day-ahead distributionally robust scheduling. IEEE Trans. Sustain. Energy **10**(4), 1653–1663 (2019)
164. H. Qiu, B. Zhao, W. Gu, R. Bo, Bi-level two-stage robust optimal scheduling for AC/DC hybrid multi-microgrids. IEEE Trans. Smart Grid **9**(5), 5455–5466 (2018)
165. B. Zhao, H. Qiu, R. Qin, X. Zhang, W. Gu, C. Wang, Robust optimal dispatch of AC/DC hybrid microgrids considering generation and load uncertainties and energy storage loss. IEEE Trans. Power Syst. **33**(6), 5945–5957 (2018)
166. A. Maulik, D. Das, Optimal power dispatch considering load and renewable generation uncertainties in an AC-DC hybrid microgrid. IET Gener., Transm. Distrib. **13**(7), 1164–1176 (2019)
167. J. Wu, X. Guan, A stochastic matching mechanism for wind generation dispatch and load shedding allocation in microgrid, in *ISGT 2014*, pp. 1–5 (2014)
168. M. Zaery, E.M. Ahmed, M. Orabi, M. Youssef, Operational cost reduction based on distributed adaptive droop control technique in DC microgrids, in *2017 IEEE Energy Conversion Congress and Exposition (ECCE)*, pp. 2638–2644 (2017)
169. W. Su, S. Yu, H. Li, H. Iu, T. Fernando, An MPC-based dual-solver optimization method for DC microgrids with simultaneous consideration of operation cost and power loss. IEEE Trans. Power Syst. 1–1 (2020)

170. R.A. Jabr, B.C. Pal, Intermittent wind generation in optimal power flow dispatching. IET Gener., Transm. Distrib. **3**(8), 66–74 (2009)
171. M. Dong, L. Li, Y. Nie, D. Song, J. Yang, Stability analysis of a novel distributed secondary control considering communication delay in DC microgrids. IEEE Trans. Smart Grid **10**(6), 6690–6700 (2019)
172. M. Belkhayat, R. Cooley, A. Witulski, Large signal stability criteria for distributed systems with constant power loads, in *Power Electronics Specialists Conference, 1995. PESC '95 Record., 26th Annual IEEE*, vol. 2, pp. 1333–1338 (1995)
173. S. Sanchez, M. Molinas, Assessment of a stability analysis tool for constant power loads in DC-grids, in *2012 15th International Power Electronics and Motion Control Conference (EPE/PEMC)*, pp. DS3b.2–1–DS3b.2–5 (2012)
174. R.K. Brayton, J.K. Moser, A theory of nonlinear networks–i. Q. Appl. Math. **22**(1), 1–33 (1964)
175. D. Karimipour, F.R. Salmasi, Stability analysis of AC microgrids with constant power loads based on Popov's absolute stability criterion. IEEE Trans. Circuits Syst. II: Express Briefs **62**(7), 696–700 (2015)
176. A.M. Rahimi, G.A. Williamson, A. Emadi, Loop-cancellation technique: a novel nonlinear feedback to overcome the destabilizing effect of constant-power loads. IEEE Trans. Veh. Technol. **59**(2), 650–661 (2010)
177. C.H. Rivetta, A. Emadi, G.A. Williamson, R. Jayabalan, B. Fahimi, Analysis and control of a buck DC-DC converter operating with constant power load in sea and undersea vehicles. IEEE Trans. Ind. Appl. **42**(2), 559–572 (2006)
178. H. Li, Y. Zhao, Z. Guo, C. Liu, T.Q. Zheng, The accurate modeling and stability analysis for flyback converter based on describing function method, in *2018 IEEE 27th International Symposium on Industrial Electronics (ISIE)*, pp. 1207–1212 (2018)
179. H. Li, F. Ren, J. Shang, B. Zhang, J. Lü, H. Qi, A novel large-signal stability analysis approach based on semi-tensor product of matrices with Lyapunov stability theorem for DC-DC converters, in *2016 IEEE Energy Conversion Congress and Exposition (ECCE)*, pp. 1–5 (2016)
180. P. Magne, B. Nahid-Mobarakeh, S. Pierfederici, DC-link voltage large signal stabilization and transient control using a virtual capacitor, in *2010 IEEE Industry Applications Society Annual Meeting*, pp. 1–8 (2010)
181. L. Xinquan, L. Gang, G. Jianping, Y. Cao, Stability analysis of C-C converter sing behavioral modeling technique, vol. 2, pp. 995–998 (2005)
182. S.R. Huddy, J.D. Skufca, Amplitude death solutions for stabilization of DC microgrids with instantaneous constant-power loads. IEEE Trans. Power Electron. **28**(1), 247–253 (2013)
183. J.G. Ciezki, R.W. Ashton, The design of stabilizing controls for shipboard DC-to-DC buck choppers using feedback linearization techniques, in *PESC 98 Record. 29th Annual IEEE Power Electronics Specialists Conference (Cat. No.98CH36196)*, vol. 1, pp. 335–341 (1998)
184. I. Kondratiev, E. Santi, R. Dougal, G. Veselov, Synergetic control for DC-DC buck converters with constant power load, in *2004 IEEE 35th Annual Power Electronics Specialists Conference (IEEE Cat. No.04CH37551)*, vol. 5, pp. 3758–3764 (2004)
185. M. Cupelli, M. Moghimi, A. Riccobono, A. Monti, A comparison between synergetic control and feedback linearization for stabilizing MVDC microgrids with constant power load, in *IEEE PES Innovative Smart Grid Technologies, Europe*, pp. 1–6 (2014)
186. M. Srinivasan, A. Kwasinski, Decentralized control of a vehicular microgrid with constant power loads, in *2014 IEEE International Electric Vehicle Conference (IEVC)*, pp. 1–8 (2014)
187. A. Kwasinski, P.T. Krein, Stabilization of constant power loads in DC-DC converters using passivity-based control, in *INTELEC 07 - 29th International Telecommunications Energy Conference*, pp. 867–874 (2007)
188. J. Samanes, A. Urtasun, E.L. Barrios, D. Lumbreras, J. López, E. Gubia, P. Sanchis, Control design and stability analysis of power converters: the MIMO generalized bode criterion. IEEE J. Emerg. Sel. Top. Power Electron. **8**(2), 1880–1893 (2020)
189. Y. Li, L. Fan, Stability analysis of two parallel converters with voltage-current droop control. IEEE Trans. Power Deliv. **32**(6), 2389–2397 (2017)
190. Y. Guo, L. Chen, X. Lu, J. Wang, T. Zheng, S. Mei, Region-based stability analysis for active dampers in AC microgrids. IEEE Trans. Ind. Appl. **55**(6), 7671–7682 (2019)

191. H. Li, F. Ren, C. Liu, Z. Guo, J. Lü, B. Zhang, T.Q. Zheng, An extended stability analysis method for paralleled DC-DC converters system with considering the periodic disturbance based on Floquet theory. IEEE Access **8**, 9023–9036 (2020)
192. A. Elbkosh, D. Giaouris, V. Pickert, B. Zahawi, S. Banerjee, Stability analysis and control of bifurcations of parallel connected DC/DC converters using the monodromy matrix, in *2008 IEEE International Symposium on Circuits and Systems (ISCAS)*, pp. 556–559 (2008)
193. J. Yan, Y. Qi, Stability analysis of a parallel-converter system with master/slave configuration, in *2019 IEEE 58th Conference on Decision and Control (CDC)*, pp. 8136–8141 (2019)
194. J.M. Zhang, X.G. Xie, X.K. Wu, Z. Qian, Stability study for paralleled DC/DC converters, in *2004 IEEE 35th Annual Power Electronics Specialists Conference (IEEE Cat. No.04CH37551)*, vol. 2, pp. 1569–1575 (2004)
195. F. Guo, Z. Lian, C. Wen, Q. Xu, Decentralized communication-free secondary voltage restoration and current sharing control for islanded DC microgrids, in *IECON 2019 - 45th Annual Conference of the IEEE Industrial Electronics Society*, vol. 1, pp. 6515–6520 (2019)
196. Y. Han, X. Ning, P. Yang, L. Xu, Review of power sharing, voltage restoration and stabilization techniques in hierarchical controlled DC microgrids. IEEE Access **7**, 149202–149223 (2019)
197. J. Nsengiyaremye, B.C. Pal, M.M. Begovic, Microgrid protection using low-cost communication systems. IEEE Trans. Power Deliv. **35**(4), 2011–2020 (2020)
198. A. Hooshyar, R. Iravani, Microgrid protection. Proc. IEEE **105**(7), 1332–1353 (2017)
199. Z.C. Li, J. Hu, J.H. He, T. Yip, A protection method for microgrids based on information sharing, in *12th IET International Conference on Developments in Power System Protection (DPSP 2014)*, pp. 1–5 (2014)
200. S. Gopalan, V. Sreeram, H. Iu, Y. Mishra, An improved protection strategy for microgrids, in *IEEE PES ISGT Europe 2013*, pp. 1–5 (2013)
201. A.H.M. Soares, J.C.M. Vieira, Case study: adaptative overcurrent protection scheme applied to an industrial plant with cogeneration units, in *2008 IEEE/PES Transmission and Distribution Conference and Exposition*, pp. 1–5 (2008)
202. J.H. He, Y.H. Cheng, J. Hu, H.T. Yip, An accelerated adaptive overcurrent protection for distribution networks with high DG penetration, in *13th International Conference on Development in Power System Protection 2016 (DPSP)*, pp. 1–5 (2016)
203. Y.G. Mostafa, M.S. Aly, Neural network based overcurrent voltage controlled protection system in large electrical networks, in *2009 IEEE Bucharest PowerTech*, pp. 1–6 (2009)
204. Y. Xue, J.M. Guerrero, Smart inverters for utility and industry applications, in *Proceedings of PCIM Europe 2015; International Exhibition and Conference for Power Electronics, Intelligent Motion, Renewable Energy and Energy Management*, pp. 1–8 (2015)
205. N. Bottrell, T.C. Green, Comparison of current-limiting strategies during fault ride-through of inverters to prevent latch-up and wind-up. IEEE Trans. Power Electron. **29**(7), 3786–3797 (2014)
206. Y. Ueda, K. Kurokawa, T. Tanabe, K. Kitamura, H. Sugihara, Analysis results of output power loss due to the grid voltage rise in grid-connected photovoltaic power generation systems. IEEE Trans. Ind. Electron. **55**(7), 2744–2751 (2008)
207. R. Tonkoski, L.A.C. Lopes, T.H.M. El-Fouly, Coordinated active power curtailment of grid connected PV inverters for overvoltage prevention. IEEE Trans. Sustain. Energy **2**(2), 139–147 (2011)
208. P.D.F. Ferreira, P.M.S. Carvalho, L.A.F.M. Ferreira, M.D. Ilic, Distributed energy resources integration challenges in low-voltage networks: voltage control limitations and risk of cascading. IEEE Trans. Sustain. Energy **4**(1), 82–88 (2013)
209. P.N. Vovos, A.E. Kiprakis, A.R. Wallace, G.P. Harrison, Centralized and distributed voltage control: impact on distributed generation penetration. IEEE Trans. Power Syst. **22**(1), 476–483 (2007)
210. H. Chen, P.D.S. Assala, Y. Cai, P. Yang, Intelligent transient overvoltages location in distribution systems using wavelet packet decomposition and general regression neural networks. IEEE Trans. Ind. Inform. **12**(5), 1726–1735 (2016)

Chapter 3
Notations and Theoretical Preliminaries

In this chapter, several notations are introduced that are being used throughout the entire thesis. In addition, a general overview of theoretical preliminaries is given, which is paramount for following and understanding the analysis developed in the following chapters.

3.1 Vector, Matrix and Function Notations

Let $\mathbf{1}_n$, $\mathbf{1}_{n\times n}$ and $\mathbf{0}_n$, $\mathbf{0}_{n\times n}$ denote the n-dimensional vector and matrix, respectively, of unit and zero entries, and let $\mathbf{1}_n^{\perp}$ be the orthogonal complement of $\mathbf{1}_n$ in $\mathbb{R}^n$, that is, $\mathbf{1}_n^{\perp} \triangleq \{x \in \mathbb{R}^n : x \perp \mathbf{1}_n\}$. Consider I_n denoting the identity unit matrix. Given an n-tuple $(x_1, \ldots, x_n)$, let $\mathbf{x} \in \mathbb{R}^n$ be the associated vector. For an ordered index set $\mathcal{I}$ of cardinality $\mid \mathcal{I} \mid$ and an one-dimensional array $\{x_i\}_{i\in\mathcal{I}}$, we define $[\mathbf{x}] = diag\left(\{x_i\}_{i\in\mathcal{I}}\right) \in \mathbb{R}^{|\mathcal{I}|\times|\mathcal{I}|}$ to be the associated diagonal matrix.

For $x \in \mathbb{R}^n$, define the vector-valued and matrix-valued, respectively, functions $\mathbf{sin}\,(x) = (sin(x_1), \ldots, sin\,(x_n)), \mathbf{cos}\,(x) = (cos(x_1), \ldots, cos\,(x_n))$ and $[\mathbf{sin}\,(x)] = diag\left(\{sin(x_i)\}_{i\in\mathcal{I}}\right)$, $[\mathbf{cos}\,(x)] = diag\left(\{cos(x_i)\}_{i\in\mathcal{I}}\right)$.

3.2 Linear Matrix Analysis

A square matrix A is said to be a *Hermitian matrix* whenever $A = A^*$, i.e. $a_{ij} = \overline{a}_{ji}$, which * represents the complex analog of symmetry.

Lemma 1 *With $\lambda_1 \leq \lambda_2 \leq \cdots \leq \lambda_n$ representing the eigenvalues of a Hermitian matrix A, and $\beta_1 \leq \beta_2 \leq \cdots \leq \beta_n$ the eigenvalues of a Hermitian matrix B, let it hold that*

A.-C. Braitor, *Advanced Hierarchical Control and Stability Analysis of DC Microgrids*, Springer Theses, https://doi.org/10.1007/978-3-030-95415-4_3

$$\lambda_i + \beta_1 \leq \eta_i \leq \lambda_i + \beta_n \tag{3.1}$$

where $\eta_1 \leq \eta_2 \leq \cdots \leq \eta_n$ *are the eigenvalues of the Hermitian matrix* $A + B$.

Proof Presented in [1, Chap. 7]. □

A *unitary matrix* is defined as a complex square matrix $U^{n\times n}$ whose columns (or rows) constitute an orthonormal basis for $\mathbb{C}^n$. Similarly, if the matrix is real, whose columns (or rows) constitute an orthonormal basis for $\mathbb{R}^n$, then it represents an *orthogonal* matrix. Both, unitary and orthogonal matrices, have useful features, one of which is the fact that inverting them is straightforward since $U^{-1} = U^*$, with * denoting the conjugate transpose.

Lemma 2 *Let* $Q, R \in \mathbb{C}^{n\times n}$ *be two unitary matrices, i.e.* $Q^*Q = I_n$ *and* $R^*R = I_n$. *Then* $P = QR$ *is also an unitary matrix.*

Proof By calculating the product P^*P, it is shown that

$$P^*P = (QR)^*QR = R^*\left(Q^*Q\right)R = R^*R = I_n \tag{3.2}$$

which completes the proof. □

Consider matrix A real symmetric, then the inertia of matrix A is defined to be the triple (ρ, ν, ζ), where ρ, ν, and ζ represent the number of positive, negative and null eigenvalues, respectively, counting algebraic multiplicities.

Lemma 3 (Sylvester's Law of Inertia) *With* A *and* B *real symmetric matrices, let* $A \sim B$ *denote that matrices* A *and* B *are congruent, i.e.* $C^*AC = B$ *for some non-singular matrix* C. *Sylvester's law of inertia states that:*

$$A \sim B \tag{3.3}$$

if and only if A *and* B *have the same inertia.*

Proof Presented in [1, Chap. 7].[1] □

Lemma 4 *Let matrix* S *be a positive-semidefinite Hermitian and matrix* D *a positive-definite Hermitian. Then*

(1) Matrix product SD *(or* DS*) is diagonalisable. There exists a unitarily diagonalisable matrix* M *similar to* SD.
(2) If $S, D \in \mathbf{R}^{n\times n}$, *the eigenvalues of* SD *(or* DS*) have only real part, and the product* SD *(or* DS*) has the same number of negative (zero, or positive) eigenvalues as matrix* A.

---

[1] Inertia being invariant under congruence is also a corollary of a deeper theorem stating that the eigenvalues of A vary continuously with the matrix entries. The argument follows from the perturbation of the eigenvalues of the nonsingular matrix A (otherwise using $A + \epsilon I$, for small ϵ), and expressing matrix C using the $\mathbb{QR}$ factorisation.

Proof Matrix SD is similar to the symmetric matrix $D^{\frac{1}{2}}(SD)D^{-\frac{1}{2}} = D^{\frac{1}{2}}SD^{\frac{1}{2}}$, hence it is diagonalisable. By employing polar decomposition SD (or DS) can be written as $SD = UP$, where U is unitary and $P = \sqrt{(SD)^* SD}$ is a unique positive-semidefinite Hermitian matrix. Define Q to satisfy $Q^2 = U$ as in [2, Chap. 12]. Note that $M = Q^{-1}(SD)Q = QPQ$ is Hermitian, hence, by eigendecomposition $M = V\Lambda V^{-1}$, with V unitary and Λ diagonal with the eigenvalues of M (and the same index of inertia as SD) as main diagonal entries. It can be concluded that $(QV)^{-1}SD(QV) = \Lambda$, with QV unitary according to Lemma 2. This proves conclusion (1). As the similarity transformation $D^{\frac{1}{2}}(SD)D^{-\frac{1}{2}} = D^{\frac{1}{2}}SD^{\frac{1}{2}}$ is congruent to S; then, according to Lemma 3, SD has the same index of inertia as matrix S. □

For cases when one deals with large matrices, splitting them into blocks can be very convenient when aiming to compute determinants. The theorem below introduces a way of computing determinants for 2 × 2 block matrices based on the commutativity of its block elements.

Theorem 1 *If* $M = \begin{bmatrix} A & B \\ C & D \end{bmatrix}$, *where* $A, B, C, D \in \mathbb{C}^{n \times n}$, *and*

$$if\ CD = DC \quad then\ det\,M = det\,(AD - BC)\,; \tag{3.4}$$

$$if\ AC = CA \quad then\ det\,M = det\,(AD - CB)\,; \tag{3.5}$$

$$if\ BD = DB \quad then\ det\,M = det\,(DA - BC)\,; \tag{3.6}$$

$$if\ AB = AB \quad then\ det\,M = det\,(DA - CB)\,. \tag{3.7}$$

If all four blocks of matrix M commute pairwise, then Eqs. (3.4)–(3.7) *are equivalent, with the determinants being equal.*

Proof Presented in [3, Chap. 4]. □

Lemma 5 (Quadratic Eigenvalue Problem) *Consider the matrix polynomial*

$$Q(\lambda) = \lambda^2 M + \lambda C + K \tag{3.8}$$

and matrices M, C, and K Hermitian of size n, for a quadratic eigenvalue problem (QEP) as

$$|\lambda^2 M + \lambda C + K| = 0 \tag{3.9}$$

If M, C, and K are all positive definite, then $Re(\lambda) < 0$.

Proof An extensive proof, considering the full or semi definiteness of the matrix coefficients is given in [4]. □

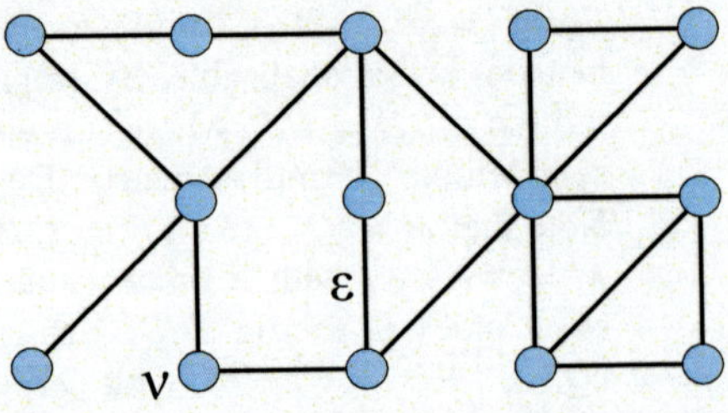

Fig. 3.1 An undirected, connected and unweighted graph $\mathcal{G}$

3.3 Elements of Graph Theory

Consider the undirected, connected, and unweighted graph $\mathcal{G}(\mathcal{V}, \mathcal{E}, A)$, represented in Fig. 3.1 as a set of vertices $\mathcal{V} = [\nu_1\ \nu_2 \ldots \nu_n]$ connected by a set of edges $\mathcal{E} \subset \mathcal{V} \times \mathcal{V}$, and induced by the symmetrical, irreducible, and nonnegative adjacency matrix $A = [a_{ij}] \in \mathbb{R}^{n\times n}$, with n being the number of vertices. The elements of A represent the weights, where $a_{ij} > 0$ if the edge $(\nu_j, \nu_i) \in \mathcal{E}$, otherwise, $a_{ij} = 0$. Here, the matrix A is assumed to be time-invariant. The Laplacian matrix $\mathcal{L} \in \mathbb{R}^{n\times n}$ is defined as $\mathcal{L} = diag\left(\left\{\sum_{j=1}^{n} a_{ij}\right\}_{i=1}^{n}\right) - A$, and its eigenvalues determine the global dynamics. For a connected graph, there is one spanning tree, with $ker\,(\mathcal{L}) = span\,(1_n)$, having all $n - 1$ remaining eigenvalues of $\mathcal{L}$ real and strictly positive, with the second-smallest eigenvalue $\lambda_2\,(\mathcal{L})$ called the *algebraic connectivity*.

It is natural for one to wonder if the intermediate eigenvalues of a Hermitian matrix have representations similar to those for the extreme eigenvalues as described in Lemma 1. The answer was given in the early years of the 20th century for not only matrices, but infinite-dimensional operators as well.

Theorem 2 (Courant-Fischer Formula) *Let A be a Hermitian matrix having the eigenvalues $\lambda_1 \le \lambda_2 \le \cdots \le \lambda_n$ with the corresponding eigenvectors $v_1, v_2, \ldots, v_n$. Then, the eigenvalues can be expressed as follows:*

$$\lambda_1 = \min_{\|x\|=1} x^T Ax = \min_{x\neq 0} \frac{x^T Az}{x^T x}, \tag{3.10}$$

$$\lambda_2 = \min_{\|x\|=1} x^T Ax = \min_{\substack{x\neq 0\\ x\perp v_1}} \frac{x^T Az}{x^T x}, \tag{3.11}$$

$$\vdots$$

$$\lambda_n = \lambda_{max} = \max_{\|x\|=1} x^T Ax = \max_{x\neq 0} \frac{x^T Ax}{x^T x}. \tag{3.12}$$

For $k \in \mathbb{Z}$ satisfying the inequality $1 \le k \le n$, let S_k denote the span of $v_1, v_2, \ldots, v_n$, and $S_k^{\perp}$ denote the orthogonal complement of S_k. Then

$$\lambda_k = \min_{\substack{||x||=1 \\ x \in S_{k-1}^{\perp}}} x^T A x = \min_{\substack{x \neq 0 \\ x \in S_{k-1}^{\perp}}} \frac{x^T A x}{x^T x}. \tag{3.13}$$

Proof Presented in [1, Chap. 7]. □

A useful application of the Courant-Fisher Formula to the Laplacian of a graph is the Rayleigh Quotient.

Corollary 1 (Rayleigh Quotient) *Consider $\mathcal{L}$ the Laplacian of the graph $\mathcal{G} = (\mathcal{V}, \mathcal{E})$. One knows that the smallest eigenvalues of a graph is $\lambda_1 = 0$ with the eigenvector $v_1 = \mathbf{1}_n$. By virtue of Theorem 1,*

$$\lambda_2 = \min_{\substack{x \neq 0 \\ x \perp v_1}} x^T A x = \min_{\substack{x \neq 0 \\ x \perp \mathbf{1}_n}} \frac{\sum_{(i,j)\in\mathcal{E}} \left(x_i - x_j\right)^2}{\sum_{i\in\mathcal{V}} x_i^2}, \tag{3.14}$$

$$\vdots$$

$$\lambda_{max} = \max_{x \neq 0} x^T A x = \max_{x \neq 0} \frac{\sum_{(i,j)\in\mathcal{E}} \left(x_i - x_j\right)^2}{\sum_{i\in\mathcal{V}} x_i^2}. \tag{3.15}$$

One can interpret the formula for λ_2, introduced in Corollary 1, similar to putting strings on each edge and minimizing the potential energy of the configuration.

3.4 Nonlinear System Analysis

Let the system

$$\dot{x} = f(t, x) \tag{3.16}$$

where $f : [0, \infty) \times \mathbb{D} \to \mathbb{R}^n$ is piecewise continuous in t and locally Lipschitz in x on $[0, \infty) \times \mathbb{D}$, and $\mathbb{D} \subset \mathbb{R}^n$ is a domain that contains the origin.

Consider the following Lyapunov-like theorem for showing *uniform boundedness* and *ultimate boundedness*.

Theorem 3 (Ultimate boundedness) *Let $\mathbb{D} \subset \mathbb{R}^n$ be a domain that contains the origin and $V : [0, \infty) \times \mathbb{D} \to \mathbb{R}$ be a continuously differentiable functions such that*

$$\alpha_1 (||x||) \leq V(t, x) \leq \alpha_2 (||x||) \tag{3.17}$$

$$\frac{\partial V}{\partial t} + \frac{\partial V}{\partial x} f(t, x) \leq -W_3(x), \quad \forall\, ||x|| \geq \mu > 0 \tag{3.18}$$

$\forall t \geq 0$ and $\forall x \in \mathbb{D}$, where α_1 and α_2 are class $\mathcal{K}$ functions and $W_3(x)$ is a continuous positive definite function. Take $r > 0$ such that $B_r \subset \mathbb{D}$ and suppose that

$$\mu < \alpha_2^{-1}\left(\alpha_1(r)\right) \tag{3.19}$$

then, there exists a class $\mathcal{KL}$ function β and for every initial state $x(t_0)$, satisfying $\|x(t_0)\| \leq \alpha_2^{-1}(\alpha_1(r))$, there is $T \geq 0$ (dependent on $x(t_0)$ and μ) such that the solution of (3.16) *satisfies*

$$\|x(t)\| \leq \beta\left(\|x(t_0)\|, t - t_0\right), \quad \forall\, t_0 \leq t \leq t_0 + T \tag{3.20}$$

$$\|x(t)\| \leq \alpha_1^{-1}\left(\alpha_2(\mu)\right), \quad \forall\, t \geq t_0 + T \tag{3.21}$$

Moreover, if $\mathbb{D} = \mathbb{R}^n$ and α_1 belongs to the class $\mathcal{K}_\infty$, then (3.20) *and* (3.21) *hold for any initial state $x(t_0)$, with no restriction on how large μ is.*

Proof Presented in [5, Theorem 4.18]. □

Inequalities (3.20) and (3.21) show that $x(t)$ is uniformly bounded for all $t \geq t_0$ and uniformly ultimately bounded with the ultimate bound $\alpha_1^{-1}(\alpha_2(\mu))$, which is a class $\mathcal{K}$ function of μ. Therefore, the smaller the value of μ, the smaller the ultimate bound.

One of the main applications of this theorem emerges from studying the stability of perturbed systems.

Theorem 4 (Singular perturbation problem) *Consider the singularly perturbed system*

$$\dot{x} = f(t, x, z, \varepsilon) \tag{3.22}$$

$$\epsilon\dot{z} = g(t, x, z, \varepsilon) \tag{3.23}$$

Assume that the following assumptions are satisfied for all

$$(t, x, \varepsilon) \in [0, \infty) \times B_r \times [0, \varepsilon] \tag{3.24}$$

- *$f(t, 0, 0, \varepsilon) = 0$ and $g(t, 0, 0, \varepsilon) = 0$.*
- *The equation*

$$0 = g(t, x, z, 0) \tag{3.25}$$

  has an isolated root $z = h(t, x)$ such that $h(t, 0) = 0$.
- *The functions f, g, h, and their partial derivatives up to the second order are bounded for $z - h(t, x) \in B_\rho$.*
- *The origin of the reduced system*

$$\dot{x} = f(t, x, h(t, x), 0) \tag{3.26}$$

  is exponentially stable.

- *The origin of the boundary-layer system*

$$\frac{dy}{d\tau} = g\left(t, x, y + h\left(t, x\right), 0\right) \tag{3.27}$$

is exponentially stable, uniformly in (t, x).

Then, there exists $\varepsilon^* > 0$ *such that for all* $\varepsilon < \varepsilon^*$, *the origin of* (3.22)–(3.23) *is exponentially stable.*

Proof Presented in [5, Theorem 11.4]. □

References

1. C.D. Meyer, *Matrix Analysis and Applied Linear Algebra* (Society for Industrial and Applied Mathematics, Philadelphia, 2000)
2. S.R.Garcia, R.A. Horn, *A Second Course in Linear Algebra*. Cambridge Mathematical Textbooks (Cambridge University Press, Cambridge, 2017)
3. E. Artin, *Geometric Algebra* (Interscience, 1957)
4. F. Tisseur, K. Meerbergen, The quadratic eigenvalue problem. SIAM Rev. **43**, 08 (2001)
5. H.K.Khalil, *Nonlinear Systems*, 3rd edn. (Pearson, 2014)

Chapter 4
Admittance Matrix Computation and Stability Analysis of Droop-Controlled DC Microgrids

In this chapter, a different approach towards computing and factorising the admittance matrix is proposed. This methodology of manipulating the admittance matrix facilitates and ensures the acquisition of stability conditions for droop-controlled DC microgrids with CILs, CCLs and/or CPLs. Such strategy is particularly useful when a generic controller (droop control, voltage regulation) is designed to control the terminal DC bus voltage of the microgrid, because it permits the factorisation of the admittance matrix by separating the singular matrices. This guarantees a simpler approach towards proving closed-loop stability by isolating singularities, whilst by employing fundamental linear algebra tools one arrives at sufficient stability conditions. To verify the proposed approach, compute the admittance matrix and test the stability conditions, a DC microgrid is considered consisting of n DC/DC converter units supplying a common CPL. Simulation testing is performed to prove the desired operation of the controller and assess the control design performance.

4.1 DC Microgrid System Modelling

The system under consideration is a generic DC microgrid, depicted in Fig. 4.1, consisting of n power converter units connected in a parallel configuration to a common DC bus that feeds an equivalent load.

One can notice, that the output currents have the following expression

$$i_i = \frac{V_i - V_o}{R_i}, \quad i \in \mathcal{I} \tag{4.1}$$

A.-C. Braitor, *Advanced Hierarchical Control and Stability Analysis of DC Microgrids*, Springer Theses, https://doi.org/10.1007/978-3-030-95415-4_4

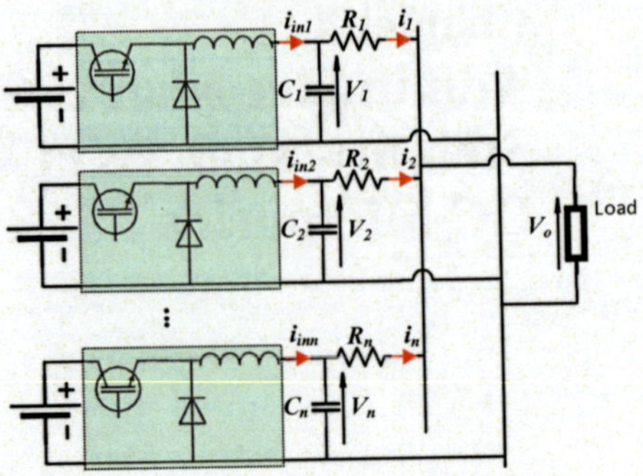

Fig. 4.1 Typical reduced-model architecture of a DC microgrid

Thus, the load voltage can be derived as follows depending on the load type:

1. constant impedance (Z) load: The characteristic equation can be written as

$$\sum_{i=1}^{n} \frac{V_i - V_o}{R_i} = V_o G, \tag{4.2}$$

where $\frac{1}{G}$ is constant and represents the load resistance. From (4.2), the load voltage becomes

$$V_o = \frac{\sum_{i=1}^{n} \frac{V_i}{R_i}}{G + \sum_{i=1}^{n} \frac{1}{R_i}}. \tag{4.3}$$

2. constant current (I) load: In this case the characteristic equation is

$$\sum_{i=1}^{n} \frac{V_i - V_o}{R_i} = i_{load}, \tag{4.4}$$

where i_{load} is constant and represents the load current. The expression of the load voltage becomes

$$V_o = \frac{\sum_{i=1}^{n} \frac{V_i}{R_i} - i_{load}}{\sum_{i=1}^{n} \frac{1}{R_i}}. \tag{4.5}$$

3. constant power (P) load: The power balance equation yields

$$V_o \sum_{i=1}^{n} \frac{V_i - V_o}{R_i} = P, \tag{4.6}$$

where P is constant and represents the power of the P load.

The expression of the load voltage will be given by the solutions of a second-order polynomial. Consider now the following assumption:

Assumption 1 For $\forall\, i = \mathcal{I}$, it holds that

$$\left(\sum_{i=1}^{n} \frac{V_i}{R_i}\right)^2 > 4P \sum_{i=1}^{n} \frac{1}{R_i}.$$

Thus, the load voltage expression is given by the real solutions in the following form

$$V_o = \frac{\sum_{i=1}^{n} \frac{V_i}{R_i} \pm \sqrt{\left(\sum_{i=1}^{n} \frac{V_i}{R_i}\right)^2 - 4P \sum_{i=1}^{n} \frac{1}{R_i}}}{2 \sum_{i=1}^{n} \frac{1}{R_i}}. \tag{4.7}$$

The load voltage (4.7) has two solutions, a high voltage and a low voltage, with the high voltage representing the feasible solution, a fact also considered in [1, 2]. Therefore

$$V_o = \frac{\sum_{i=1}^{n} \frac{V_i}{R_i} + \sqrt{\left(\sum_{i=1}^{n} \frac{V_i}{R_i}\right)^2 - 4P \sum_{i=1}^{n} \frac{1}{R_i}}}{2 \sum_{i=1}^{n} \frac{1}{R_i}}. \tag{4.8}$$

Hence, a generalised expression for the load voltage in all three load cases can be found as

$$V_o = \frac{\sum_{i=1}^{n} \frac{V_i}{R_i} + \alpha}{\beta + \sum_{i=1}^{n} \frac{1}{R_i}}. \tag{4.9}$$

where α and β have the expressions specified in Table 4.1.

In Fig. 4.2, the reduced microgrid is depicted when considering a fast inner current control loop. By applying Kirchhoff's laws, the governing dynamic equations of the capacitor voltages, in matrix form, are the following

$$C\dot{V} = i_{in} - i, \tag{4.10}$$

where $C = diag\{C_i\}$, $V = [V_1 \ldots V_n]$ represents the state vector capacitor voltages, $i = [i_1 \ldots i_n]$ is the output current vector, and $i_{in} = [i_{in1} \ldots i_{inn}]$ is the control input vector representing the current of each converter.

Table 4.1 Parameters α and β for ZIP load

	Z	I	P
α	0	$-i_{load}$	$\sqrt{\left(\sum_{i=1}^{n} \frac{V_i}{R_i}\right)^2 - 4P \sum_{i=1}^{n} \frac{1}{R_i}}$
β	G	0	$\sum_{i=1}^{n} \frac{1}{R_i}$

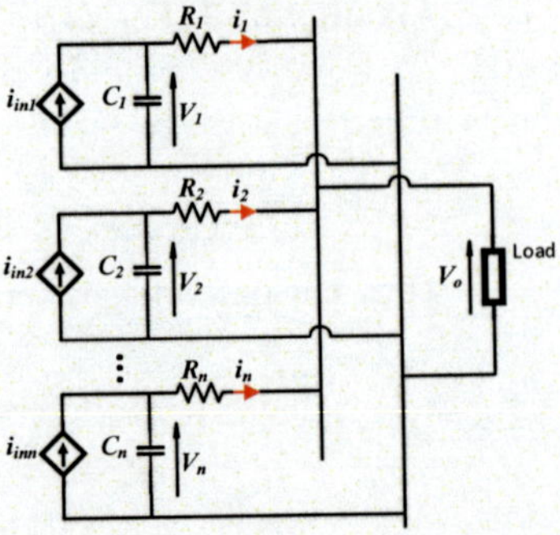

Fig. 4.2 Reduced-model of the DC system considering a fast inner current control loop

Remark 4 System (4.10) represents a generic model of n-sourced units that could be integrated with the microgrid via different power converter configurations (buck, boost, buck-boost, AC/DC) in the given DC microgrid framework.

4.2 Admittance Matrix

By taking the partial derivative of the output current i_i from (4.1) with respect to the capacitor voltage V_i, it yields the admittance matrix

$$\begin{aligned}
Y = \frac{\partial i}{\partial V} &= \begin{bmatrix} \frac{1}{R_1}\left(1-\frac{\partial V_o}{\partial V_1}\right) & -\frac{1}{R_1}\frac{\partial V_o}{\partial V_2} & \cdots & -\frac{1}{R_1}\frac{\partial V_o}{\partial V_n} \\ -\frac{1}{R_2}\frac{\partial V_o}{\partial V_1} & \frac{1}{R_2}\left(1-\frac{\partial V_o}{\partial V_2}\right) & \cdots & -\frac{1}{R_2}\frac{\partial V_o}{\partial V_n} \\ \vdots & \vdots & \ddots & \vdots \\ -\frac{1}{R_n}\frac{\partial V_o}{\partial V_1} & -\frac{1}{R_n}\frac{\partial V_o}{\partial V_2} & \cdots & \frac{1}{R_n}\left(1-\frac{\partial V_o}{\partial V_n}\right) \end{bmatrix} \\
&= \begin{bmatrix} \frac{1}{R_1} & 0 & \cdots & 0 \\ 0 & \frac{1}{R_2} & \cdots & 0 \\ \vdots & \vdots & \ddots & \vdots \\ 0 & 0 & \cdots & \frac{1}{R_n} \end{bmatrix} \left(I_n - \begin{bmatrix} \frac{\partial V_o}{\partial V_1} & \frac{\partial V_o}{\partial V_1} & \cdots & \frac{\partial V_o}{\partial V_1} \\ \frac{\partial V_o}{\partial V_2} & \frac{\partial V_o}{\partial V_2} & \cdots & \frac{\partial V_o}{\partial V_2} \\ \vdots & \vdots & \ddots & \vdots \\ \frac{\partial V_o}{\partial V_n} & \frac{\partial V_o}{\partial V_n} & \cdots & \frac{\partial V_o}{\partial V_n} \end{bmatrix}^T \right) \\
&= \begin{bmatrix} \frac{1}{R_1} & 0 & \cdots & 0 \\ 0 & \frac{1}{R_2} & \cdots & 0 \\ \vdots & \vdots & \ddots & \vdots \\ 0 & 0 & \cdots & \frac{1}{R_n} \end{bmatrix} \left(I_n - \mathbf{1}_{n\times n} \begin{bmatrix} \frac{\partial V_o}{\partial V_1} & 0 & \cdots & 0 \\ 0 & \frac{\partial V_o}{\partial V_2} & \cdots & 0 \\ \vdots & \vdots & \ddots & \vdots \\ 0 & 0 & \cdots & \frac{\partial V_o}{\partial V_n} \end{bmatrix} \right) \\
&= R^{-1}\left(I_n - \mathbf{1}_{n\times n} D\right)
\end{aligned} \tag{4.11}$$

where $R = diag\{R_i\}$ and $D = diag\{\frac{\partial V_o}{\partial V_i}\}$ with the following expression:

$$D = diag\left\{\frac{\partial V_o}{\partial V_i}\right\} = \begin{bmatrix} \frac{\partial V_o}{\partial V_1} & \dots & 0 \\ \vdots & \ddots & \vdots \\ 0 & \dots & \frac{\partial V_o}{\partial V_n} \end{bmatrix} =$$

$$= \frac{1}{\beta + \sum_{i=1}^{n} \frac{1}{R_i}} \left(R^{-1} + \begin{bmatrix} \frac{\partial \alpha}{\partial V_i} & \dots & 0 \\ \vdots & \ddots & \vdots \\ 0 & \dots & \frac{\partial \alpha}{\partial V_i} \end{bmatrix} \right) \tag{4.12}$$

with $\beta \geq 0$. One can note that for the Z and I load cases, there is $\begin{bmatrix} \frac{\partial \alpha}{\partial V_1} & \dots & 0 \\ \vdots & \ddots & \vdots \\ 0 & \dots & \frac{\partial \alpha}{\partial V_n} \end{bmatrix} =$ $\mathbf{0}_{n \times n}$, while for the P load case, the matrix becomes

$$diag\left\{\frac{\partial \alpha}{\partial V_i}\right\} = \begin{bmatrix} \frac{\partial \alpha}{\partial V_1} & \dots & 0 \\ \vdots & \ddots & \vdots \\ 0 & \dots & \frac{\partial \alpha}{\partial V_n} \end{bmatrix} =$$

$$= R^{-1} + \frac{\sum_{i=1}^{n} \frac{V_i}{R_i}}{\sqrt{\left(\sum_{i=1}^{n} \frac{V_i}{R_i}\right)^2 - 4P \sum_{i=1}^{n} \frac{1}{R_i}}} \begin{bmatrix} \frac{1}{R_1} & \dots & 0 \\ \vdots & \ddots & \vdots \\ 0 & \dots & \frac{1}{R_n} \end{bmatrix}$$

$$= \frac{\sum_{i=1}^{n} \frac{V_{ie}}{R_i}}{\alpha_e} R^{-1}. \tag{4.13}$$

where α_e is given from Table 4.1 with $V_i = V_{ie}$ [3]. It is clear, by virtue of Lemma 1, that the diagonal matrix D is positive-definite for the Z and I load cases, but also for the P load case given $\sqrt{\left(\sum_{i=1}^{n} \frac{V_i}{R_i}\right)^2 - 4P \sum_{i=1}^{n} \frac{1}{R_k}} > 0$ holds, according to Assumption 1. The diagonal entries and also the eigenvalues of matrix D for the P load case, as in [4], will take the form

$$\lambda_{Di} = \frac{1}{2\sum_{i=1}^{n} \frac{1}{R_i}} \left(\frac{1}{R_i} + \frac{\sum_{i=1}^{n} \frac{V_i}{R_i}}{\sqrt{\left(\sum_{i=1}^{n} \frac{V_i}{R_i}\right)^2 - 4P \sum_{i=1}^{n} \frac{1}{R_i}}} \frac{1}{R_i} \right), \tag{4.14}$$

with $i \in \mathcal{I}$.

Remark 5 In [1, 5], the power balance equation is linearised and then an expression for V_o is obtained with respect to its steady-state equilibrium, V_{oe}. This new expression of V_o is substituted in Eq. (4.1) and used to, finally, compute the admittance matrix as a function of the equilibrium point, V_{oe}, of the load voltage. On the contrary, this proposed method considers the instantaneous nonlinear expressions of the output currents, i_i from (4.1), and the load voltage V_o from (4.9), to compute the admittance matrix for every V_i. When the admittance matrix is required at a particular equilibrium point, then it can be calculated with $V_i = V_{ie}$, where V_{ie} is the value of the capacitor voltage at each node i at the equilibrium point.

4.3 Stability of Droop Controlled Microgrids

Due to the instability effect that the nonlinear load (P load, in this case) introduces, the stability proof is far from trivial. In this work, the parallel DER units in the microgrid under consideration feed a common load connected to the main bus. The main task is to achieve load voltage regulation close to a desired reference value and share the load proportionally to the sources capacities. This can be achieved through droop control operation as explained below.

4.3.1 Droop Control Design

Conventional droop controllers introduce a static structure and regulate the sources output voltage, which leads to significant load voltage drop and inaccurate power sharing. To improve the load voltage regulation and power sharing, a dynamic droop controller is introduced with the following expression

$$\dot{V} = V^*\mathbf{1}_n - V_o\mathbf{1}_n - mi, \tag{4.15}$$

with V^* being the reference voltage, and $m = diag\{m_i\}$ the droop coefficients. At the steady-state there is

$$V_o = V^* - m_i i_i, \tag{4.16}$$

which ensures accurate power sharing

$$m_1 i_1 = m_2 i_2 = \cdots = m_n i_n \tag{4.17}$$

proportionally to the sources capacities with suitable choice of m_i.

The droop controller is implemented using a proportional integral (PI) controller, in matrix form, as follows

$$i_{in} = -k_P V + \sigma \tag{4.18}$$

$$\dot{\sigma} = k_I\left(V^*\mathbf{1}_n - V_o\mathbf{1}_n - mi\right), \tag{4.19}$$

where $k_P = diag\{k_{Pi}\}$ and $k_I = diag\{k_{Ii}\}$ are the proportional and integral gains of the PI controller, respectively, and for which

$$m_i < R_i, \tag{4.20}$$

$\forall\, i \in \mathcal{I}$. By replacing the controller dynamics (4.18)–(4.19) into the open-loop system (4.10), the closed-loop system becomes

$$C\dot{V} = -k_P V + \sigma - i \tag{4.21}$$

$$\dot{\sigma} = k_I \left(V^* \mathbf{1}_n - V_o \mathbf{1}_n - mi \right) \tag{4.22}$$

where the output current i is linked to the capacitor voltage V through the impedance matrix.

4.3.2 Stability Analysis

Consider an equilibrium point (V_e, σ_e) of the closed-loop system (4.21)–(4.22), (4.1) and (4.8), satisfying Assumption 1. Then the following theorem can be formulated that guarantees stability of the entire droop-controlled DC microgrid with a CPL.

Theorem 5 *The equilibrium point (V_e, σ_e) is asymptotically stable if the following conditions holds*

$$k_{Pi} > \frac{n\lambda_{Di} - 1}{R_i}, \quad \forall\, i \in \mathcal{I}. \tag{4.23}$$

Proof The Jacobian matrix corresponding to system (4.21)–(4.22) takes the following form

$$J = \begin{bmatrix} -C^{-1}k_P - C^{-1}Y & C^{-1} \\ -k_I \mathbf{1}_{n\times n} D - k_I m Y & \mathbf{0}_{n\times n} \end{bmatrix}.$$

Replacing the admittance matrix with its expression from (4.11), it yields

$$J = \begin{bmatrix} -C^{-1}k_P - C^{-1}R^{-1}\left(I_n - \mathbf{1}_{n\times n} D\right) & C^{-1} \\ -k_I \mathbf{1}_{n\times n} D - k_I m R^{-1}\left(I_n - \mathbf{1}_{n\times n} D\right) & \mathbf{0}_{n\times n} \end{bmatrix}.$$

According to Theorem 1, the characteristic polynomial of the system can be written as

$$|\lambda I_{2n} - J| = |\lambda^2 I_n + \mathbf{C}\lambda + \mathbf{K}| = 0 \tag{4.24}$$

with

$$\mathbf{C} = C^{-1}k_P + C^{-1}R^{-1}\left(I_n - \mathbf{1}_{n\times n} D\right)$$

$$\mathbf{K} = C^{-1}\left(k_I m R^{-1} + k_I \left(I_n - mR^{-1}\right)\mathbf{1}_{n\times n} D\right)$$

By right multiplying (4.24) with $|D^{-1}| > 0$, the obtained determinant is

$$|\lambda^2 D^{-1} + \overline{\mathbf{C}}\lambda + \overline{\mathbf{K}}| = 0 \tag{4.25}$$

with

$$\overline{\mathbf{C}} = C^{-1} k_P D^{-1} + C^{-1} R^{-1} \left(D^{-1} - \mathbf{1}_{n\times n}\right)$$

$$\overline{\mathbf{K}} = C^{-1} \left(k_I m R^{-1} D^{-1} + k_I \left(I_n - m R^{-1}\right) \mathbf{1}_{n\times n}\right)$$

By left multiplying (4.25) with $|CR| = |RC| > 0$, one obtains

$$|\lambda^2 CRD^{-1} + \mathbf{C}^*\lambda + \mathbf{K}^*| = 0 \tag{4.26}$$

with

$$\mathbf{C}^* = Rk_P D^{-1} + D^{-1} - \mathbf{1}_{n\times n}$$

being a symmetric matrix and, following factorisation,

$$\mathbf{K}^* = Rk_I\left(I_n - mR^{-1}\right)\left(\left(I_n - mR^{-1}\right)^{-1} R^{-1} m D^{-1} + \mathbf{1}_{n\times n}\right)$$

being a diagonalisable matrix whose eigenvalues are all real, according to Lemma 4. Hence, by using $\mathbf{K}^* = P^{-1}\Lambda P$, where P is unitary and Λ is diagonal, the characteristic polynomial becomes

$$|\ \lambda^2 CRD^{-1} + \mathbf{C}^*\lambda + P^{-1}\Lambda P\ | = 0 \tag{4.27}$$

and, by left and right multiplication with $|\ P\ |$ and $|\ P^{-1}\ |$, respectively, it becomes

$$|\ \lambda^2 PCRD^{-1}P^{-1} + P\mathbf{C}^*P^{-1}\lambda + \Lambda\ | = 0 \tag{4.28}$$

Note that Λ is a diagonal matrix with the same index of inertia as $\mathbf{K}^*$, and the similarity transformations $PCRD^{-1}P^{-1}$ and $P\mathbf{C}^*P^{-1}$ are symmetric, as P is unitary $\left(P^{-1} = P^T\right)$, and they share the same eigenvalues as CRD^{-1}, and $\mathbf{C}^*$, respectively. If CRD^{-1}, $\mathbf{C}^*$ and Λ are positive-definite, then $Re\{\lambda\} < 0$, which means that matrix J is Hurwitz.

As CRD^{-1} is already positive-definite, it would suffice to show that $\mathbf{C}^* > 0$, and $\Lambda > 0$, or equivalently that $\mathbf{K}^*$ has positive eigenvalues.

1. Condition $\mathbf{C}^* > 0$:

$$\mathbf{C}^* = Rk_P D^{-1} + D^{-1} - \mathbf{1}_{n\times n} > 0 \tag{4.29}$$

Since $\mathbf{C}^*$ is a sum of symmetric matrices, then according to Lemma 1, condition (4.29) can be rewritten in scalar form as

$$\frac{R_i k_{Pi} + 1}{\lambda_{Di}} - n > 0 \tag{4.30}$$

which is always true, provided that (4.23) is satisfied.

2. Condition $\Lambda > 0$, or $\mathbf{K}^*$ has positive eigenvalues:
According to (4.20), the first matrix in the multiplication inside $\mathbf{K}^*$ is positive-definite, i.e.

$$Rk_I\left(I_n - mR^{-1}\right) > 0. \tag{4.31}$$

Hence, according to Lemma 2, it would suffice to investigate only the remaining symmetric matrix in the product. This matrix

$$R^{-1}\left(I_n - mR^{-1}\right)^{-1} mD^{-1} + \mathbf{1}_{n\times n} > 0 \tag{4.32}$$

is positive-definite since it is a sum between a positive-definite diagonal matrix and a positive semi-definite symmetric matrix.

Hence, when (4.23) is satisfied, J is Hurwitz, and the equilibrium point (V_e, σ_e) is asymptotically stable. This completes the proof of Theorem 5. □

4.4 Simulation Results

A DC MG has been considered, consisting of five parallel-operated buck converters, as depicted in Fig. 4.4, with arbitrary parameters given in Table 4.2 selected for demonstration purposes only. The devices are feeding a common CPL, and have been

Table 4.2 System and control parameters of a DC microgrid consisting of five DC/DC buck converters feeding a common CPL

System parameters	Values	System parameters	Values
C_1	110 μF	R_1	1 Ω
C_2	150 μF	R_2	1.1 Ω
C_3	100 μF	R_3	1.05 Ω
C_4	420 μF	R_4	1.12 Ω
C_5	200 μF	R_5	1.15 Ω
Control parameters	Values	Control parameters	Values
m_1	0.42	m_5	0.084
m_2	0.21	$k_{P1\ldots5}$	0.01
m_3	0.14	$k_{I1\ldots5}$	2×10^3
m_4	0.105	V^*	100

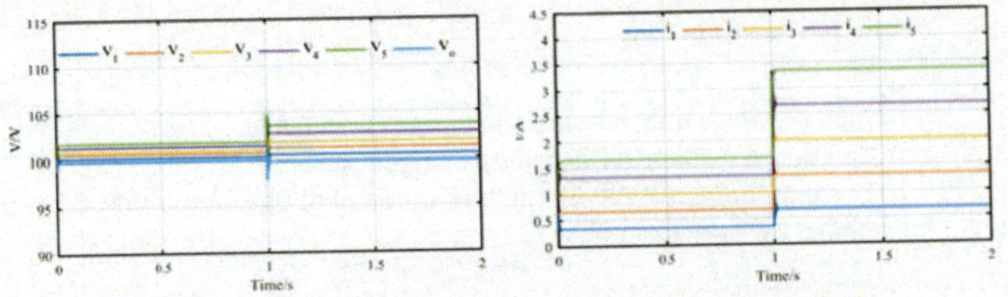

(a) Capacitor and load voltages (b) Output currents

Fig. 4.3 Simulation results of the DC microgrid system with PI controller

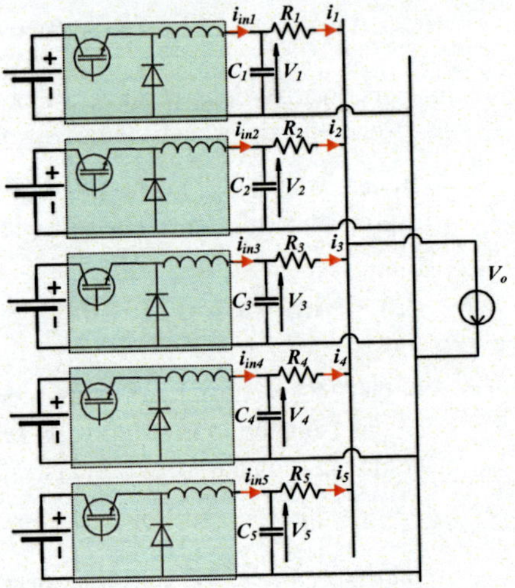

Fig. 4.4 DC microgrid considered for testing

simulated in Matlab/Simulink for 2 s using the average model of the buck converters. The proportional terms $k_{P1...5}$ have been carefully chosen to satisfy inequality (4.23).

Each converter is equipped with the droop controller considered in Sect. 4.3.2, and the task is to regulate the load voltage to the rated value, $V^* = 100\,\text{V}$, and share their output power in a $1:2:3:4:5$ ratio.

At $t = 0\,\text{s}$, the load power is $P = 500\,\text{W}$, and as one can see in Fig. 4.3a, the load voltage is accurately fixed at $V_o = 99.9\,\text{V}$. The output currents satisfy their control imposed ratios, having $i = [1.67\ \ 1.34\ \ 0.99\ \ 0.67\ \ 0.34]\,\text{A}$ (Fig. 4.3b).

After one second, at $t = 1\,\text{s}$, the load power changes to $2P = 1\,\text{kW}$. The load voltage regulation is still fairly accurate, $V_o = 99.7\,\text{V}$ (Fig. 4.3a), and, according to Fig. 4.3b, the output currents respect their assigned proportions, having $i = [3.34\ \ 2.67\ \ 2.02\ \ 1.34\ \ 0.66]\,\text{A}$.

4.5 Conclusions

A novel approach to compute the admittance matrix has been proposed that enables and facilitates the stability analysis, especially in the case where the controller aims to regulate the converter output voltage. Following the proposed strategy, after several factorisations, the isolation of singularities will be possible, and henceforth, by employing straightforward linear algebra tools, the stability conditions will be comfortably acquired. By selecting appropriate control parameters, a DC microgrid consisting of five parallel-operated DC/DC buck converters feeding a common CPL has been used for testing and validating the proposed analysis. The results confirm that the presented strategy ensures normal operation of the DC network and guarantees an elementary path in ensuring closed-loop stability.

References

1. M. Su, Z. Liu, Y. Sun, H. Han, X. Hou, Stability analysis and stabilization methods of DC microgrid with multiple parallel-connected DC-DC converters loaded by CPLs. IEEE Trans. Smart Grid **9**(1), 132–142 (2018)
2. J.W. Simpson-Porco, F. Dörfler, F. Bullo, Voltage stabilization in microgrids via quadratic droop control. IEEE Trans. Autom. Control **62**(3), 1239–1253 (2017)
3. A.C. Braitor, G.C. Konstantopoulos, V. Kadirkamanathan, Current-limiting droop control design and stability analysis for paralleled boost converters in DC microgrids. IEEE Trans. Control Syst. Technol. **29**(1), 385–394 (2020)
4. A.-C. Braitor, G.C. Konstantopoulos, V. Kadirkamanathan, Admittance matrix computation and stability analysis of droop controlled DC micro-grids with CPL. IFAC-PapersOnLine **53**(2), 13525–13530 (2020). 21st IFAC World Congress
5. Z. Liu, S. Mei, Y. Sun, H. Han, X. Hou, J.M. Guerrero, Stability analysis of DC microgrids with constant power load under distributed control methods. Automatica **90**, 62–72 (2018)

Chapter 5
Control Design and Stability Analysis of DC Microgrids Consisting of Unidirectional DC/DC Boost Converters

Motivated by the urge to design controllers that ensure system protection against overcurrents, which has attracted a lot of attention in the recent years, in this chapter, a nonlinear current-limiting droop controller is developed to guarantee load power distribution between parallel operated DC/DC boost converters in a DC microgrid application. In Sect. 5.1, the already mentioned robust droop control strategy is adopted and implemented as a dynamic virtual resistance in series with the inductance of each converter unit. Unlike traditional approaches that use small-signal modelling, the proposed control design considers the accurate nonlinear dynamic model of each converter unit and it is analytically demonstrated that accurate power sharing can be accomplished with an inherent current limitation for each converter using ultimate boundedness theory.

In Sect. 5.2, a modified current-limiting droop controller is introduced, for paralleled DC/DC boost converters loaded by ZIP loads in a DC microgrid architecture. With an improved version of the previously proposed nonlinear current-limiting controller, an inherent current-limiting property is guaranteed for each converter independently of the load type or magnitude variations. Sufficient conditions that guarantee closed-loop stability of the entire DC system with Z, I, or P loads are analytically acquired. To validate the benefits of the proposed framework, simulation and experimental results are provided for a DC microgrid consisting of paralleled DC/DC boost converters feeding either a Z, I, or P load.

5.1 Paralleled DC/DC Boost Converters Feeding a CIL

In Fig. 5.1, a typical islanded DC microgrid is depicted consisting of DC/DC power converters connected in parallel to a common DC bus and feeding a load. Power sharing without the need of communication among the different converters is often achieved via droop control [1, 2]. In the conventional droop control strategy, each

A.-C. Braitor, *Advanced Hierarchical Control and Stability Analysis of DC Microgrids*, Springer Theses, https://doi.org/10.1007/978-3-030-95415-4_5

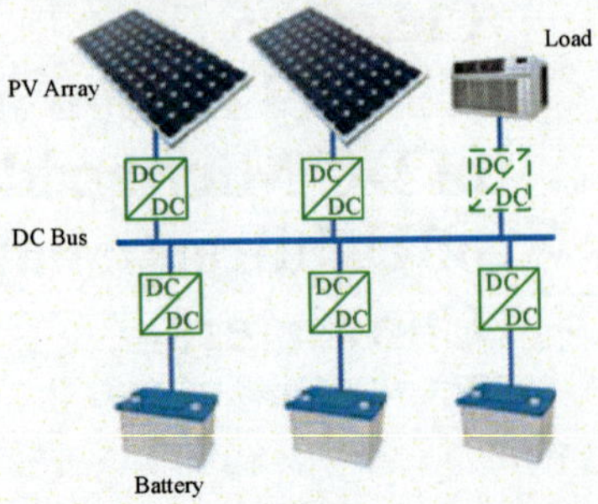

Fig. 5.1 DC microgrid topology consisting of paralleled DC/DC converters

one of the n parallel-operated power converters introduces an output voltage V_i of the form:

$$V_i = V^* - m_i i_i, \tag{5.1}$$

where i_i is the output current of each converter, m_i is the droop coefficient and $i \in \mathcal{I}$. However, conventional droop control suffers from poor voltage regulation and cannot achieve accurate power sharing when each converter introduces a different output impedance [3, 4]. One of the recently developed methods to address these issues is based on a robust droop strategy, which achieves accurate power sharing and tight voltage regulation [3, 4]. The robust droop controller takes the form

$$\dot{V}_i = k_e(V^* - V_o) - m_i i_i, \tag{5.2}$$

where V_o is the load voltage and k_e is a constant gain. At the steady state, there is

$$m_1 i_1 = m_2 i_2 = \ldots = m_n i_n. \tag{5.3}$$

By multiplying this expression with the load voltage V_o in each part of the equation, it yields

$$m_1 P_1 = m_2 P_2 = \ldots = m_n P_n, \tag{5.4}$$

where $P_i = V_o i_i$ is the power at the output injected to the load by the i-th converter. Equality (5.4) indicates the power sharing among the paralleled units.

Although accurate power sharing is achieved independently from the power requested by the load, the technical limitations of each converter are not taken into account. Given the power rating P_n of a converter and the rated output voltage V^*, a limitation is introduced for the output current (and consequently the input current) of each converter. To ensure protection to the generating circuit or the transmission system from harmful effects in cases of significant changes in the load demand, a current-limiting property is required. Hence, imposing an upper limit for the cur-

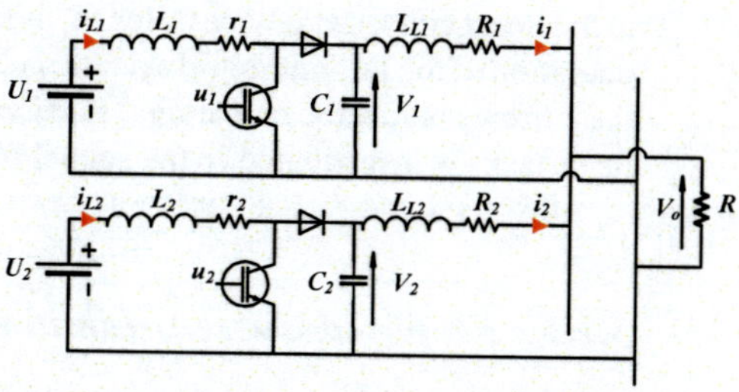

Fig. 5.2 Proposed network configuration for parallel operation

rent that may be delivered to a load and making sure that certain boundaries are not violated represents another major challenge in a DC microgrid operation.

Figure 5.2 shows the configuration of a DC microgrid consisting of two DC/DC boost converters connected in parallel and feeding a common load, which is assumed as resistive. Although for simplicity, the investigation is restricted in two paralleled converters, it can be easily expanded to the cases of n boost converters in a DC microgrid.

Using Kirchhoff laws and average analysis [5], the dynamic model of the entire system becomes

$$L_1\dot{i}_{L1} = U_1 - r_1 i_{L1} - (1-u_1)V_1 \tag{5.5}$$

$$C_1\dot{V}_1 = (1-u_1)i_{L1} - i_1 \tag{5.6}$$

$$L_{L1}\dot{i}_1 = V_1 - (i_1+i_2)R - R_1 i_1 \tag{5.7}$$

$$L_2\dot{i}_{L2} = U_2 - r_2 i_{L2} - (1-u_2)V_2 \tag{5.8}$$

$$C_2\dot{V}_2 = (1-u_2)i_{L2} - i_2 \tag{5.9}$$

$$L_{L2}\dot{i}_2 = V_2 - (i_1+i_2)R - R_2 i_2. \tag{5.10}$$

Here L_1, L_2 are the boost converter inductances with parasitic resistances r_1 and r_2, respectively, and C_1, C_2 represent the output capacitors the converters. The output impedances or line impedances of the converters are introduced by the inductances L_{L1}, L_{L2} and the resistances R_1, R_2, while R is the common load. The state vector of the system consists of the inductor currents i_{L1}, i_{L2} at the input of every converter, the output voltages V_1, V_2 and line currents i_1, i_2. The control input vector consists of the duty-ratio inputs of each converter u_1 and u_2, which by definition should remain bounded in the set [0, 1]. The DC input voltages of the converters are given as U_1 and U_2, and represent constant inputs for the system (uncontrollable), as shown in Fig. 5.2.

It can be observed, that system (5.5)–(5.10) is nonlinear, since the control inputs u_1 and u_2 are multiplied with the system states. In addition, in the case where $u_1 = 1$ or $u_2 = 1$, at the steady-state, the inductor currents i_{L1} and i_{L2} take the values $i_{L1} = \frac{U_1}{r_1}$ and $i_{L2} = \frac{U_2}{r_2}$, respectively. Since r_1 and r_2 are parasitic resistances and therefore very small, then the two input currents reach very high values that can cause damage to the boost converter devices. Hence, there is a clear challenge to achieve the desired operation of the DC microgrid system, i.e. accurate power sharing, while maintaining the currents below the converters' rated values. Such a controller that can achieve all these tasks is investigated in the sequel [6].

5.1.1 *Proposed Current-Limiting Droop Controller*

5.1.1.1 Controller Design and Analysis

In order to achieve the desired power sharing and voltage regulation, whilst keeping a limited current for each boost converter, the robust droop control concept given in (5.2) is implemented as a dynamic virtual resistance for each converter, opposed to the original design which is applied directly to the voltage. Hence, the duty-ratio input of each boost converter takes the form

$$u_i = 1 - \frac{w_i}{V_i} i_i, \tag{5.11}$$

where $i = \{1, 2\}$ indicates the converter number and w_i represents a virtual resistance for i-th converter. In order to incorporate the robust droop control concept, the virtual resistance is proposed to follow the nonlinear dynamics:

$$\dot{w}_i = -c_i w_{qi}^2 \left[k_e(V^* - V_o) - m_i i_i\right] \tag{5.12}$$

$$\dot{w}_{qi} = c_i \left(k_e(V^* - V_o) - m_i i_i\right) \frac{(w_i - w_{mi})w_{qi}}{\Delta w_{mi}} - k_{qi} \left(\frac{(w_i - w_{mi})^2}{\Delta w_{mi}} + w_{qi}^2 - 1\right) w_{qi}, \tag{5.13}$$

with c_i, k_{qi}, k_e, w_{mi}, Δw_{mi} being positive constants. It is highlighted that a second controller state w_{qi} is introduced to define the dynamic structure of the virtual resistance and to maintain a given bound for w_i. To further explain this, the nonlinear controller dynamics w_i and w_{qi} are investigated. Considering the following Lyapunov function candidate for system (5.12)–(5.13):

$$W_i = \frac{(w_i - w_{mi})^2}{\Delta w_{mi}^2} + w_{qi}^2, \tag{5.14}$$

then by calculating its time derivative and using the controller equations (5.12)–(5.13), it yields:

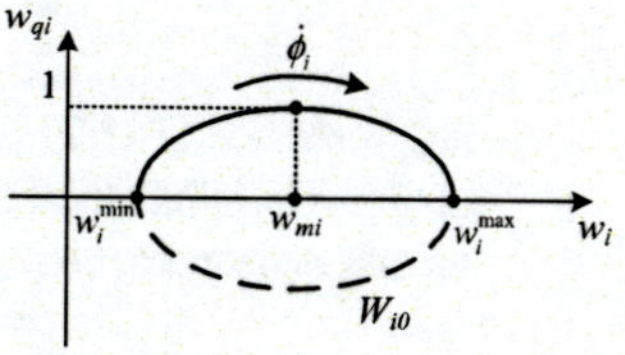

Fig. 5.3 Phase portrait of the controller dynamics

$$
\begin{aligned}
\dot{W}_i =& \frac{2\,(w_i - w_{mi})\,\dot{w}_i}{\Delta w_{mi}^2} + 2w_{qi}\dot{w}_{qi} \\
=& -2c_i w_{qi}^2 \frac{w_i - w_{mi}}{\Delta w_{mi}^2}\left(k_e(V^* - V_o) - m_i i_i\right) \\
&+ 2c_i w_{qi}^2 \frac{w_i - w_{mi}}{\Delta w_{mi}^2}\left(k_e(V^* - V_o) - m_i i_i\right) \\
&- 2k_{qi}\left(\frac{(w_i - w_{mi})^2}{\Delta w_{mi}} + w_{qi}^2 - 1\right) w_{qi}^2 \\
=& -2k_{qi}\left(\frac{(w_i - w_{mi})^2}{\Delta w_{mi}^2} + w_{qi}^2 - 1\right) w_{qi}^2.
\end{aligned}
\tag{5.15}
$$

From the expression (5.15), one can notice that $\dot{W}_i$ becomes zero on the ellipse

$$
W_{i0} = \left\{ w_i, w_{qi} \in \mathbb{R} : \frac{(w_i - w_{mi})^2}{\Delta w_{mi}^2} + w_{qi}^2 = 1 \right\}, \tag{5.16}
$$

or at the horizontal axis $w_{qi} = 0$ on the $w_i - w_{qi}$ plane (Fig. 5.3). This indicates that if the initial conditions of the controller states w_{i0} and w_{qi0} are chosen on the ellipse W_{i0}, i.e. they satisfy

$$
\frac{(w_{i0} - w_{mi})^2}{\Delta w_{mi}^2} + w_{qi0}^2 = 1 \tag{5.17}
$$

then from (5.15) there is

$$
\dot{W}_i(t) = 0,\ \forall t \geq 0, \tag{5.18}
$$

which results in

$$
W_i(t) = W_i(0) = 1,\ \forall t \geq 0, \tag{5.19}
$$

leading to the result that w_i and w_{qi} will start and remain on the ellipse W_{i0} for all $t \geq 0$, as shown in Fig. 5.3. Hence, a typical choice for the initial conditions is $w_{i0} = w_{mi},\ w_{qi0} = 1$.

Since the controller states operation is restricted on the ellipse W_{i0}, then $w_i \in [w_i^{min}, w_i^{max}] = [w_{mi} - \Delta w_{mi}, w_{mi} + \Delta w_{mi}]$ and $w_{qi} \in [0, 1]$ for all $t \geq 0$. If the positive constants w_{mi} and Δw_{mi} are chosen to guarantee

$$w_{mi} > \Delta w_{mi} \tag{5.20}$$

then $w_i^{min} > 0$, which means that the ellipse W_{i0} is located on the right-half plane of $w_i - w_{qi}$ and therefore $w_i \in [w_i^{min}, w_i^{max}] > 0, \ \forall t \geq 0$, introducing a positive virtual resistance. Using the transformation

$$w_i - w_{mi} = \Delta w_{mi} \sin\phi \tag{5.21}$$

$$w_{qi} = \cos\phi \tag{5.22}$$

inside the controller dynamics (5.12)–(5.13), after a few calculations it results in

$$\dot{\phi}_i = \frac{c_i w_{qi}^2}{\Delta w_{mi}} \left(k_e(V^* - V_o) - m_i i_i\right) \tag{5.23}$$

which proves that the controller state trajectory on the $w_i - w_{qi}$ plane will move on the ellipse W_{i0} with an angular velocity $\dot{\phi}_i$ given by (5.23). It is highlighted that the angular velocity becomes zero when:

(i) $k_e(V^* - V_o) - m_i i_i = 0$, which guarantees the accurate power sharing and the desired tight voltage regulation, or
(ii) $w_{qi} = 0$, which leads to $w_i = w_i^{min}$ or $w_i = w_i^{max}$, corresponding to the current-limiting capability as explained in the sequel.

5.1.1.2 Current Limitation

By substituting the expression of the proposed controller (5.11) into the inductor current Eqs. (5.5) and (5.8), the closed-loop dynamics of the inductor current become for each converter:

$$L_i \dot{i}_{Li} = -\left(w_i + r_i\right) i_{Li} + U_i. \tag{5.24}$$

By introducing the following Lyapunov function candidate

$$V_i = \frac{1}{2} L_i i_{Li}^2 \tag{5.25}$$

and computing its time derivative, after using (5.24), the expression of $\dot{V}$ becomes

$$\dot{V}_i = L_i i_{Li} \cdot \dot{i}_{Li} = -\left(w_i + r_i\right) i_{Li}^2 + U_i i_{Li} \tag{5.26}$$

Taking into account that $w_i \in [w_i^{min}, w_i^{max}] > 0, \ \forall t \geq 0$, as proven in the previous subsection then

$$
\begin{aligned}
\dot{V}_i &\leq -\left(w_i^{min} + r_i\right) i_{Li}^2 + U_i i_{Li} \\
&\leq -\left(w_i^{min} + r_i\right) |i_{Li}|^2 + |U_i|\, |i_{Li}|
\end{aligned} \tag{5.27}
$$

Thus

$$
\dot{V}_i < -r_i\, |i_{Li}|^2, \; \forall\, |i_{Li}| \geq \frac{U_i}{w_i^{min}}. \tag{5.28}
$$

According to (5.28), the solution $i_{Li}(t)$ is uniformly ultimately bounded (Theorem 3), and every solution starting with the initial condition $i_{Li}(0)$, satisfying

$$
|i_{Li}(0)| \leq \frac{U_i}{w_i^{min}}, \tag{5.29}
$$

will remain in this range for all future times, i.e.

$$
|i_{Li}(t)| \leq \frac{U_i}{w_i^{min}}, \; \forall t \geq 0. \tag{5.30}
$$

By selecting w_i^{min} as

$$
w_i^{min} = \frac{U_i}{i_{Li}^{max}} \tag{5.31}
$$

where i_{Li}^{max} represents the maximum input current allowed to flow through the converter according to the converter ratings, then by substituting (5.31) into (5.30), it yields

$$
|i_{Li}(t)| \leq \frac{U_i}{\frac{U_i}{i_{Li}^{max}}} = i_{Li}^{max}, \; \forall t \geq 0 \tag{5.32}
$$

which guarantees the desired current-limiting capability of each boost converter separately.

It is highlighted that the current-limiting property of each converter is guaranteed independently from the power sharing function $k_e(V^* - V_o) - m_i i_i$ than needs to be regulated to zero. This means that each converter has as the first priority to protect itself from high currents that can damage the device. When the current is below the maximum value, then power sharing can be achieved. This will be illustrated in the simulation results that follows.

5.1.2 Simulation Results

To evaluate the proposed control strategy, a DC microgrid is considered for simulation testing, consisting of two parallel DC/DC boost converters, similar to the one presented in Fig. 5.2. The microgrid is simulated using the average model of the boost

Table 5.1 Controller and system parameters of a DC microgrid consisting of two parallel-operated DC/DC boost converters feeding a common CIL

System parameters	Values	Control parameters	Values
$L_{L1,2}$	[0.2 0.21] mH	$m_{1,2}$	[1 2]
$R_{1,2}$	2.2 Ω	$c_{1,2}$	[1 2] × 10^3
$L_{1,2}$	2.2 mH	$w_{m1,2}$	[505 1040]
$r_{1,2}$	0.5 Ω	$i_{L1,2}^{min}$	100 mA
$U_{1,2}$	[200 100] V	$k_{q1,2}$	1000
$C_{1,2}$	560 μF	$i_{L1,2}^{max}$	[2.5 10] A
R	225 Ω	k_e	10
$P_{1,2}$	[0.5 1] kW	t_s	0.1 s

converters from Simpower Systems toolbox in Matlab/Simulink. The system and controller parameters displayed in Table 5.1 are selected arbitrarily for demonstration purposes where, however, given $i_{L1,2}^{min}$ and $i_{L1,2}^{max}$, one can obtain $w_{m1,2}$. The main task is to achieve accurate power sharing in a 2 : 1 ratio among the paralleled converters and tight regulation of the common load voltage to the rated value $V^* = 300$ V, while maintaining the inductor currents below their maximum values independently from the load changes. Each converter is equipped with the proposed controller.

At $t = 0$ s, the load is $R = 225\ \Omega$, and as it can be seen in Fig. 5.4b, accurate power sharing is achieved having $i_1 = 0.9$ A and $i_2 = 0.45$ A at the steady state, satisfying $i_1 = 2i_2$. Figure 5.4c illustrates the line and load voltages, and it is clear that the load voltage is regulated very close to the rated value $V^* = 300$ V, while the line voltages V_1 and V_2 are a bit higher to achieve the desired power sharing. Figure 5.4a depicts the inductor currents i_{L1} and i_{L2} which stay below the limits imposed by the system's technical requirements.

Five seconds later, at $t = 5$ s, a load change is applied and the resistive load changes to 150 Ω. One can see in Fig. 5.4c, that after a small transient, the line voltages slightly increase and the load voltage V_o remains close to the 300 V value as intended. The inductor currents and the line currents increase due to the higher power demand, but the accurate power sharing is maintained, since $i_1 = 1.33$ and $i_2 = 0.67$ at the steady state, as shown in Fig. 5.4b. The inductor currents still remain below their bounds (Fig. 5.4a).

Finally, at $t = 20$ s a second load change occurs and the resistive load becomes 90 Ω. By increasing the power demand further and requesting higher currents from each converter, as pictured in Fig. 5.4a, the inductor current of converter 1 reaches the limit $i_{L1}^{max} = 2.5$ A based on the proposed current-limiting strategy, while the inductor current of the second converter still stays below its maximum value. Therefore, power sharing is sacrificed to protect the first power converter from damages, as it is shown in Fig. 5.4b. Nevertheless, the load voltage is still regulated close to the rated value as required (Fig. 5.4c).

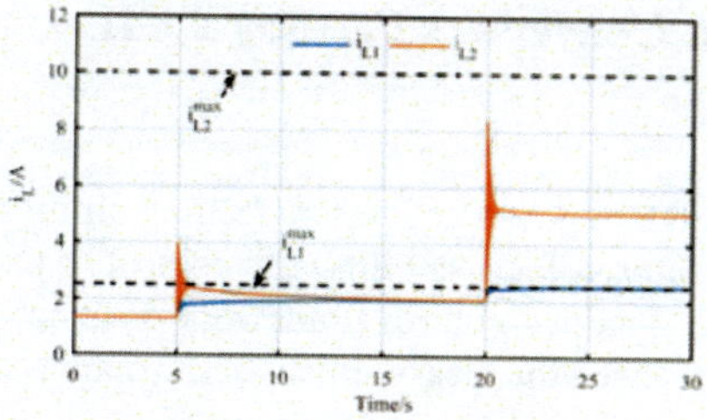

(a) inductor currents i_{L1} and i_{L2}

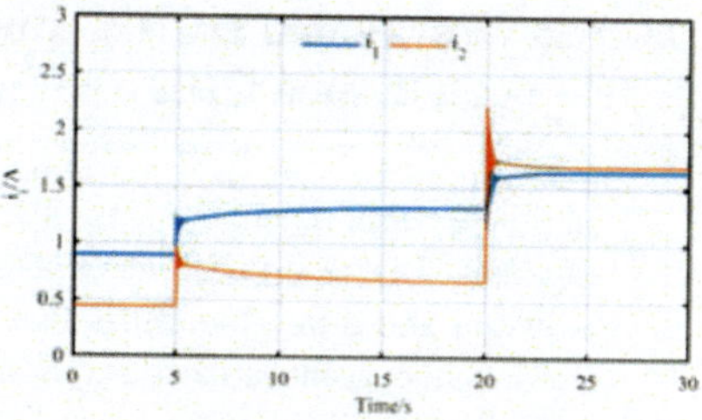

(b) line currents i_1 and i_2

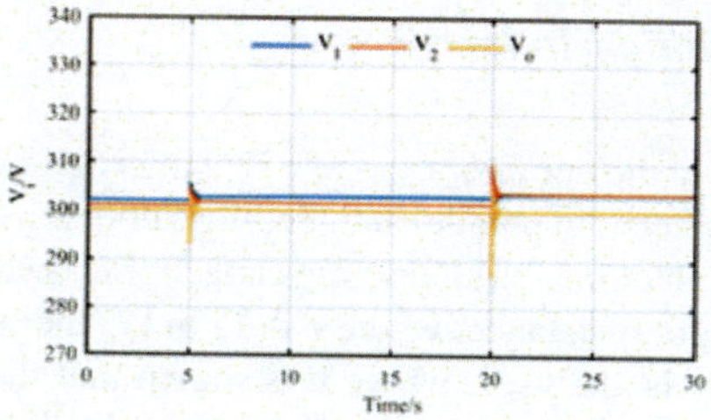

(c) load voltage V_o

Fig. 5.4 Simulation results of the system states of two parallel operated DC/DC boost converters under the proposed controller

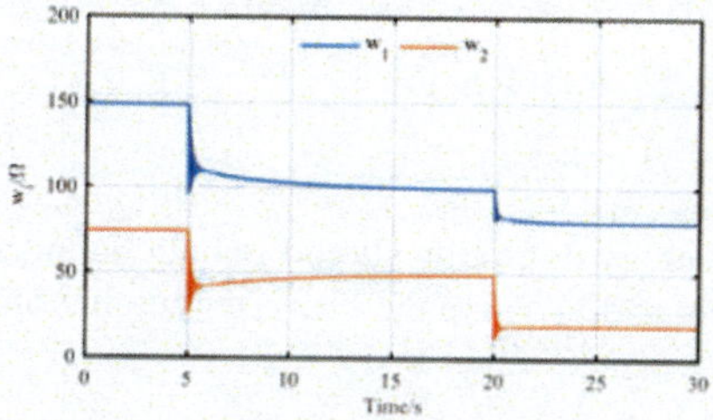

(a) virtual resistances w_1 and w_2

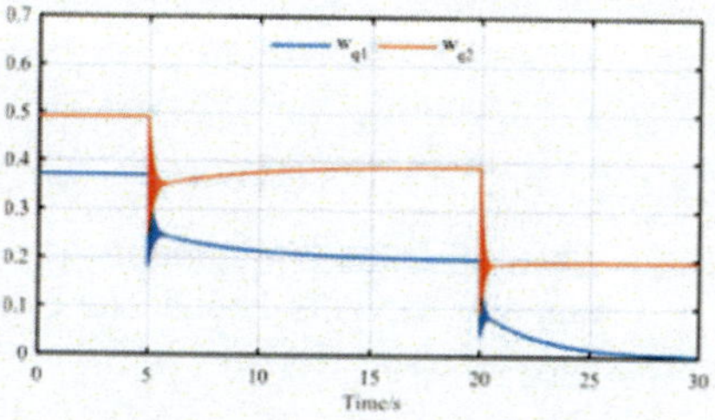

(b) controller states w_{q1} and w_{q2}

Fig. 5.5 Simulation results of the controller states of two parallel operated DC/DC boost converters under the proposed controller

The transient response of the virtual resistances is displayed in Fig. 5.5a. It is observed that as the load decreases and consequently the power demand increases, both virtual resistances decrease to allow a higher current flow. At the final change of the load, w_1 reaches its minimum value $w_1^{min} = \frac{U_1}{i_{L1}^{max}} = 80\,\Omega$ which limits the inductor current i_{L1} below its given maximum value. The response of the additional controller states w_{q1} and w_{q2} is provided in Fig. 5.5b. By combining the values of w_i and w_{qi} given in Fig. 5.5a, b, it is verified that $\frac{(w_i-w_{mi})^2}{\Delta w_{mi}^2} + w_{qi}^2 = 1$ holds true, which validates the theoretical development.

5.2 Paralleled DC/DC Boost Converters Feeding a CIL, CCL or a CPL

In this section, the analysis is extended to n power converters feeding different types of loads. In the process of ensuring closed-loop stability, the controller had to be adjusted and a key benefit is that the obtained control framework does not require additional current sensors as it does not need the measurement of the output currents.

5.2.1 Dynamic Model

Figure 5.6 shows the configuration of a DC microgrid consisting of n DC/DC boost converters connected in parallel and feeding a common load. Each converter consists of a boosting inductor L_i, a smoothing capacitor C_i, while U_i is the DC input voltage and R_i the output resistance, where $i \in \mathcal{I}$. In [7] the impact of cable impedance on system stability is analysed, where it is shown that the inductance has no effect on the system stability; hence, for simplicity the cable impedance is regarded as purely resistive.

Using Kirchhoff laws and average analysis [5], the nonlinear dynamic model of each DC/DC boost converter, can be described by the following differential equations:

$$L_i \dot{i}_{Li} = U_i - (1 - u_i)V_i \tag{5.33}$$

$$C_i \dot{V}_i = (1 - u_i)i_{Li} - i_i \tag{5.34}$$

where u_i is the duty-ratio input, which by definition should remain bounded in the range [0, 1], i_{Li} is the inductor current and V_i, i_i are the converter output voltage and current, respectively.

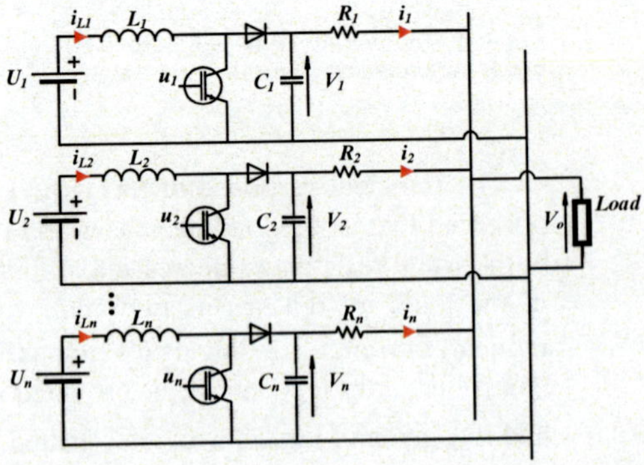

Fig. 5.6 DC microgrid configuration with n paralleled unidirectional DC/DC boost converters feeding a common generic load

Rewriting (5.33)–(5.34) in a matrix form, the DC microgrid system takes the following form

$$\dot{i}_L = L^{-1}\left(U - (I_n - u)V\right) \tag{5.35}$$

$$\dot{V} = C^{-1}\left((I_n - u)i_L - i\right) \tag{5.36}$$

where $U = [U_1 ... U_n]^T$, $u = diag\{u_i\}$, $V = [V_1 ... V_n]^T$, $i_L = [i_{L1} ... i_{Ln}]^T$, $i = [i_1 ... i_n]^T$, $L = diag\{L_i\}$ and $C = diag\{C_i\}$.

One can observe that system (5.35)–(5.36) is nonlinear, since the control input u is multiplied with the system states, $\left[i_L^T \; V^T\right]^T$.

Assumption 2 For $\forall i = \mathcal{I}$, there is $V_i \geq U_i$, which represents a requirement for any DC/DC boost converter.

Please note that based on Assumptions 1 and 2, solution (4.7) exists if

$$\left(\sum_{i=1}^{n} \frac{U_i}{R_i}\right)^2 - 4P\sum_{i=1}^{n} \frac{1}{R_i} > 0.$$

Assumption 3 Let *Assumption* 2 hold. If $i_{Li}^{max} > 0$ represents the maximum inductor current of each converter, i.e. $| i_{Li} | < i_{Li}^{max}$, let

$$U_i - i_{Li}^{max} R_i > \frac{\sum_{i=1}^{n} \frac{V_i}{R_i} - \sqrt{\left(\sum_{i=1}^{n} \frac{V_i}{R_i}\right)^2 - 4P\sum_{i=1}^{n} \frac{1}{R_i}}}{2\sum_{i=1}^{n} \frac{1}{R_i}} \tag{5.37}$$

hold.

5.2.2 *Proposed Controller Design*

The aim of the control design is to achieve power sharing among the paralleled converters and tight load voltage regulation close to the rated value, while maintaining a limited input current for each converter. Here the droop control concept is implemented as a dynamic virtual resistance for each converter, opposed to the traditional design, which is applied directly to the voltage. Hence, the duty-ratio input of each boost converter takes the form

$$u_i = 1 - \frac{w_i}{V_i} i_{Li}, \tag{5.38}$$

where $i \in \mathcal{I}$ indicates the converter number and w_i represents a virtual resistance for i-th converter. Inspired by the current-limiting controller [8], the virtual resistance is proposed to follow the nonlinear dynamics:

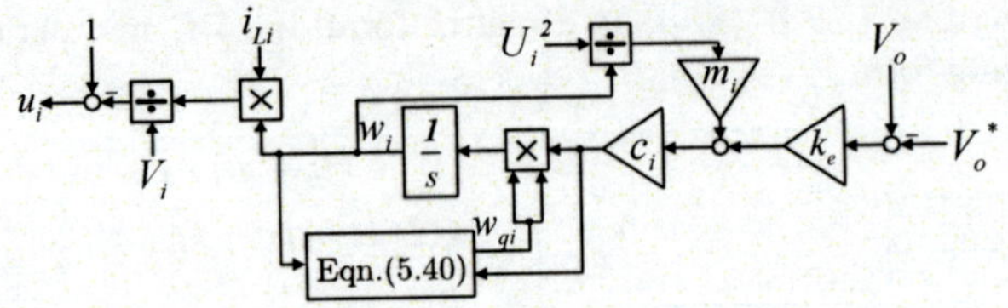

Fig. 5.7 Implementation diagram of the proposed controller (5.38)–(5.40)

$$\dot{w}_i = -c_i w_{qi}^2 \left[k_e(V_o^* - V_o) - m_i \frac{U_i^2}{w_i} \right] \tag{5.39}$$

$$\dot{w}_{qi} = c_i \frac{(w_i - w_{mi})}{\Delta w_{mi}^2} w_{qi} \left[k_e(V_o^* - V_o) - m_i \frac{U_i^2}{w_i} \right] - c_i k_{qi} \left(\frac{(w_i - w_{mi})^2}{\Delta w_{mi}^2} + w_{qi}^2 - 1 \right) w_{qi}, \tag{5.40}$$

with $c_i, k_{qi}, k_e, w_{mi}, \Delta w_{mi}$ being positive constants and V_o^*, m_i representing the load voltage reference and the droop coefficient, respectively. In contrast to the robust droop controller [3], the proposed controller does not require the measurement of the output current i_i of each converter; thus leading to a simpler implementation and also facilitating the stability analysis in Sect. 5.2.3. It is highlighted that the proposed structure of the control dynamics guarantees a given bound for w_i based on the bounded integral controller concept [9]. For more details on the boundedness of w_i and w_{qi} the reader is referred to [9] where it is shown that $w_i \in \left[w_i^{min}, w_i^{max}\right] > 0$ and $w_{qi} \in [0, 1]$ for all $t \geq 0$, given typical initial conditions $w_{i0} = w_{mi}$ and $w_{qi0}=1$. Note also that due to the current-limiting property of the proposed controller, given $i_{Li}^{max} > 0$, then if $w_i^{min} = w_{mi} - \Delta w_{mi} = \frac{U_i}{i_{Li}^{max}}$, then $| i_{Li}(t) | \leq i_{Li}^{max}$, $\forall t \geq 0$ (see [8]). A control diagram with the controller and all sensed feedback variables is shown in Fig. 5.7.

Assumption 4 For every constant $w_{ie} \in \left(w_i^{min}, w_i^{max}\right) > 0$ satisfying

$$m_1 \frac{U_1^2}{w_{1e}} = m_2 \frac{U_2^2}{w_{2e}} = \ldots = m_n \frac{U_n^2}{w_{ne}}. \tag{5.41}$$

there exists a unique steady-state equilibrium point $\left(i_{Lie}, V_{ie}, w_{ie}, w_{qie}\right)$ corresponding to the desired voltage regulation, i.e.

$$V_{oe} = V_o^* - \frac{m_i U_i^2}{k_e w_{ie}} \tag{5.42}$$

where $w_{qie} \in (0, 1]$, $\forall i = \mathcal{I}$.

Assumption 5 For $\forall i = \mathcal{I}$, it holds that $\frac{k_e}{m_i V_{ie}} - \frac{1}{R_i} > 0$.

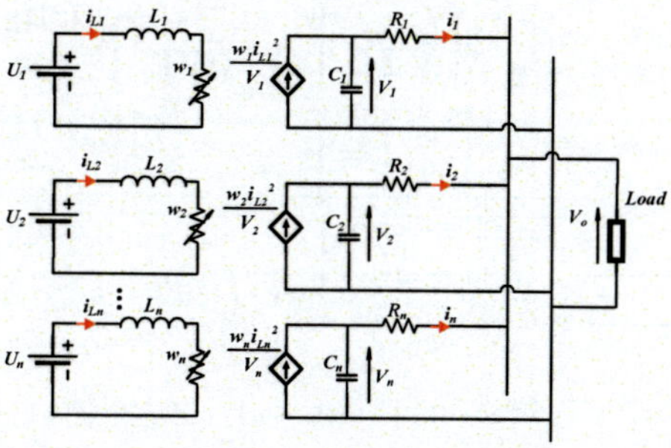

Fig. 5.8 Equivalent circuit of the closed-loop system

By replacing the expression of the proposed controller (5.38) into the inductor current Eq. (5.33), the closed-loop dynamics of the inductor current for each converter become:

$$L_i \dot{i}_{Li} = -w_i i_{Li} + U_i, \tag{5.43}$$

where it is clear that w_i represents a virtual resistance in series with the inductance L_i. The equivalent closed-loop system is given in Fig. 5.8, where it is clear that the current i_{Li} dynamics of each converter are partially decoupled from the voltages V_i. At the steady-state there is

$$i_{Lie} = \frac{U_i}{w_{ie}}. \tag{5.44}$$

Hence, the term $\frac{U_i^2}{w_i}$ represents the input power of each converter at the steady-state. As a result, (5.41) yields

$$m_1 P_1 = m_2 P_2 = ... = m_n P_n$$

which indicates the desired power sharing in the DC microgrid based on a suitable choice of the droop parameters m_i.

5.2.3 *Stability Analysis*

5.2.3.1 Closed-Loop System

By applying the proposed controller (5.38)–(5.40) into the DC microgrid dynamics (5.33)–(5.34), the closed-loop system can be written in the following matrix form

$$\begin{bmatrix} \dot{i}_L \\ \dot{V} \end{bmatrix} = \begin{bmatrix} L^{-1}\left(U - [w]\, i_L\right) \\ C^{-1}\left([V]^{-1}[i_L]^2 w - i\right) \end{bmatrix} \tag{5.45}$$

$$\begin{bmatrix} \dot{w} \\ \dot{w}_q \end{bmatrix} = \begin{bmatrix} -c\left[w_q\right]^2 \left(k_e(V_o^* - V_o)\mathbf{1}_n - [U]^2[w]^{-1} m\right) \\ c([w] - w_m)\Delta w_m^{-2}\left[w_q\right]\left(k_e(V_o^* - V_o)\mathbf{1}_n - [U]^2[w]^{-1} m\right) \end{bmatrix} - \begin{bmatrix} \mathbf{0}_n \\ ck_q\left(([w] - w_m)^2\Delta w_m^{-2} + \left[w_q\right]^2 - I_n\right) w_q \end{bmatrix} \tag{5.46}$$

where $w = [w_1 ... w_n]^T$, $w_q = \left[w_{q1} ... w_{qn}\right]^T$, $c = diag\{c_i\}$, $m = [m_1 ... m_n]^T$, $k_q = diag\{k_{qi}\}$, $w_m = ediag\{w_{mi}\}$, $\Delta w_m = diag\{\Delta w_{mi}\}$.

Consider an equilibrium point $\left(i_{Lie}, V_{ie}, w_{ie}, w_{qie}\right)$ satisfying Assumption 4. By setting $\varepsilon = \frac{1}{min\{c_i\}}$, there exists $\delta = diag\{\delta_i\} \geq 0$ such that $c = \frac{1}{\varepsilon} I_n + \delta$. Hence, (5.46) becomes

$$\begin{bmatrix} \varepsilon\dot{w} \\ \varepsilon\dot{w}_q \end{bmatrix} = \begin{bmatrix} -(I_n + \varepsilon\delta)\left[w_q\right]^2 \left(k_e(V_o^* - V_o)\mathbf{1}_n - [U]^2[w]^{-1} m\right) \\ (I_n + \varepsilon\delta)([w] - w_m)\Delta w_m^{-2}\left[w_q\right]\left(k_e(V_o^* - V_o)\mathbf{1}_n - [U]^2[w]^{-1} m\right) \end{bmatrix} - \begin{bmatrix} \mathbf{0}_n \\ (I_n + \varepsilon\delta)\, k_q\left(([w] - w_m)^2\Delta w_m^{-2} + \left[w_q\right]^2 - I_n\right) w_q \end{bmatrix} \tag{5.47}$$

Hence, the closed-loop system Eqs. (5.45) and (5.47) can be written as

$$\dot{x} = f(x, z) \tag{5.48}$$

$$\varepsilon\dot{z} = g(x, z, \varepsilon) \tag{5.49}$$

where $x = \begin{bmatrix} i_L - i_{Le} \\ V - V_e \end{bmatrix}$ and $z = \begin{bmatrix} w - w_e \\ w_q - w_{qe} \end{bmatrix}$. For arbitrarily large values of the controller gains c_i, then ε is small and therefore (5.48)–(5.49) can be investigated as a singularly perturbed system using two-time-scale analysis [10]. The controller's system (5.47) is also named as the boundary layer since it represents the immediate vicinity of a bounding surface and is first analysed in the sequel.

5.2.3.2 Boundary Layer Stability Analysis

Considering f, g being continuously differentiable in the domain $(x, z, \varepsilon) \in D_x \times D_z \times [0, \varepsilon_0]$, when the controller gain c is selected sufficiently large, then $\varepsilon \to 0$ and, based on singular perturbation theory, (5.49) results in the algebraic form of $0 = g(x, z)$ as follows

$$\begin{bmatrix} \mathbf{0}_n \\ \mathbf{0}_n \end{bmatrix} = \begin{bmatrix} -\left[w_q\right]^2 \left(k_e(V_o^* - V_o)\mathbf{1}_n - [U]^2 [w]^{-1} m\right) \\ ([w] - w_m)\Delta w_m^{-2} \left[w_q\right] \left(k_e(V_o^* - V_o)\mathbf{1}_n - [U]^2 [w]^{-1} m\right) \end{bmatrix} - \begin{bmatrix} \mathbf{0}_n \\ k_q \left(([w] - w_m)^2 \Delta w_m^{-2} + \left[w_q\right]^2 - I_n\right) w_q \end{bmatrix}. \tag{5.50}$$

The roots of the above system can be computed as

$$\begin{bmatrix} w \\ w_q \end{bmatrix} = \begin{bmatrix} \frac{1}{k_e(V_o^* - V_o)} [U]^2 m \\ \left(I_n - ([w] - w_m)^2 \Delta w_m^{-2}\right)^{1/2} \mathbf{1}_n \end{bmatrix} \tag{5.51}$$

and can also be referred to as $z = h(x)$ with $w_i \in (w_i^{min}, w_i^{max}) > 0$ and $w_{qi} \in (0, 1]$, such that $h(0) = 0$. Thus, the roots also represent the equilibrium points of the nonlinear system (5.45)–(5.46). Exponential stability at the origin can be investigated via its corresponding Jacobian matrix:

$$J_1 = \begin{bmatrix} -\left[w_{qe}\right]^2 [U]^2 [w_e]^{-2} [m] & \mathbf{0}_{n\times n} \\ -\left(([w_e] - w_m) [U]^2 [w_e]^{-2} [m] + 2k_q ([w_e] - w_m)\right) \left(\Delta w_m^{-2} \left[w_{qe}\right]\right) & -2k_q \left[w_{qe}\right]^2 \end{bmatrix}$$

Matrix J_1 is Hurwitz as it is lower triangular and all diagonal elements are negative. Hence, there exist $\rho_1 > 0$ and a domain $\tilde{D}_z = \left\{z \in R^{2n}, \| z \|_2 < \rho_1\right\}$, where $\tilde{D}_z \subseteq D_z$ such that (5.49) is exponentially stable at the origin uniformly in x.

5.2.3.3 Reduced Model

To obtain the reduced model, the roots w and w_q are substituted from (5.51) into (5.45), yielding

$$\dot{i}_L = L^{-1}\left(U - \frac{1}{k_e(V_o^* - V_o)} [U]^2 [m] i_L\right) \tag{5.52}$$

$$\dot{V} = C^{-1}\left(\frac{1}{k_e(V_o^* - V_o)} [U]^2 [V]^{-1} [i_L]^2 m - i\right). \tag{5.53}$$

This model is often referred to as quasi-steady-state model, because w, w_q introduce a velocity $\left[\dot{w}\ \dot{w}_q\right]^T = \varepsilon^{-1} g$ which is very large when ε is small and $g \neq 0$, leading to rapid convergence to a root $h(i_L, V)$, which is the equilibrium of the boundary-layer. The corresponding Jacobian matrix of the reduced system will have the following form

$$J_2 = \begin{bmatrix} -\frac{1}{k_e(V_o^* - V_{oe})} L^{-1} [U]^2 [m] & -\frac{1}{V_o^* - V_{oe}} L^{-1} [U] \mathbf{1}_{n\times n} D \\ 2C^{-1} [U] [V_e]^{-1} & C^{-1}\left(k_e [V_e]^{-1} [m]^{-1} \left(\mathbf{1}_{n\times n} D - (V_o^* - V_{oe}) [V_e]^{-1}\right) - Y\right) \end{bmatrix}$$

By virtue of Theorem 1, the characteristic equation can be calculated from

$$|\lambda I_n - J_2| = |\lambda^2 I_n + \lambda M + N| = 0, \tag{5.54}$$

with

$$M = \frac{1}{k_e\left(V_o^* - V_{oe}\right)} L^{-1}[U]^2[m] - C^{-1}\left(k_e\,[V_e]^{-1}[m]^{-1}\left(\mathbf{1}_{n\times n} D - (V_o^* - V_{oe})[V_e]^{-1}\right) - Y\right) \tag{5.55}$$

$$N = \frac{1}{k_e\left(V_o^* - V_{oe}\right)} L^{-1}[U]^2[m]\,C^{-1}\left(k_e\,[V_e]^{-1}[m]^{-1}\left((V_o^* - V_{oe})\,[V_e]^{-1} + \mathbf{1}_{n\times n} D\right) + Y\right) \tag{5.56}$$

Replacing matrix Y with its expressions from (4.11), and isolating matrix $\mathbf{1}_{n\times n}$ by factorisation [11, 12], followed by left and right multiplication with determinants $|D| > 0$ and $|X| = \frac{1}{k_e(V_o^* - V_{oe})} \mid L^{-1}[U]^2[m]\,C^{-1}\left(k_e\,[V_e]^{-1}[m]^{-1} - R^{-1}\right) \mid > 0$, according to Assumption 5, respectively, the characteristic polynomial becomes

$$\mid \lambda^2 X^{-1} D^{-1} + \lambda \overline{M} + \overline{N} \mid = 0, \tag{5.57}$$

which is a quadratic eigenvalue problem (QEP) with $\overline{N}$ symmetrical and $\overline{M}$, according to Lemma 2 in [13], diagonalisable whose eigenvalues are all real, since it is a product between a positive-definite diagonal and a symmetrical matrix. Since $\overline{M} = P^{-1}\Lambda P$, Eq. (5.57) can be rewritten as

$$\mid \lambda^2 P X^{-1} D^{-1} P^{-1} + \lambda \Lambda + P \overline{N} P^{-1} \mid = 0, \tag{5.58}$$

with Λ being a diagonal matrix with the eigenvalues of matrix $\overline{M}$ as main entries. The similarity transformation $PX^{-1}D^{-1}P^{-1}$ and $P\overline{N}P^{-1}$ are symmetrical, as P is an orthogonal matrix, i.e. $P^{-1} = P^T$, and they share the same eigenvalues as $X^{-1}D^{-1} > 0$ and $\overline{N}$, respectively. If $\Lambda > 0$, or equivalently $\overline{M}$ has positive eigenvalues, and $\overline{N} > 0$, then $Re(\lambda) < 0$ and J_2 is Hurwitz. Since matrix $\overline{M}$ is represented by a multiplication where one term is the diagonal matrix $X^{-1}\left(C^{-1}k_e\,[V_e]^{-1}[m]^{-1} + C^{-1}R^{-1}\right) > 0$, according to Sylvester's law on inertia, matrix $\overline{M}$ will have the same index of inertia as the remaining term, $\left(C^{-1}k_e\,[V_e]^{-1}[m]^{-1} + C^{-1}R^{-1}\right)^{-1}\overline{M}$. According to Lemma 1 in [13], if

$$\frac{U_i^2 m_i}{k_e\left(V_o^* - V_{oe}\right)L_i} + \frac{1}{C_i}\left(\frac{k_e}{m_i V_{ie}}\left(\frac{V_o^* - V_{oe}}{V_{ie}} - \lambda_{Di} n\right) + \frac{1}{R_i}(1 - \lambda_{Di} n)\right) > 0, \tag{5.59}$$

holds $\forall i \in \mathcal{I}$, then $\overline{M} > 0$ is satisfied. For the Z and I load cases, all eigenvalues of matrix D are

$$\lambda_{Di} = \frac{1}{\beta + \sum_{i=1}^{n} \frac{1}{R_i}} \frac{1}{R_i} > 0, \tag{5.60}$$

while for the P load case, there is

$$\lambda_{Di} = \frac{1}{\beta + \sum_{i=1}^{n} \frac{1}{R_i}} \frac{1}{R_i} \left(1 + \frac{\sum_{i=1}^{n} \frac{V_{ie}}{R_i}}{\alpha_e} \right) > 0. \tag{5.61}$$

Regarding condition $\overline{N} > 0$, by considering Assumption 5, if

$$\frac{k_e}{m_i V_{ie}} \left(\frac{V_o^* - V_{oe}}{V_{ie}\lambda_{Di}} + n \right) + \frac{1}{R_i} \left(\frac{1}{\lambda_{Di}} - n \right) > 0, \ \forall i = 1...n. \tag{5.62}$$

holds, then $\overline{N} > 0$ is satisfied. Hence, if the two conditions (5.59), (5.62) hold for each converter then there exist $\rho_2 > 0$ and a domain $\tilde{D}_x = \left\{ x \in R^{2n}, \| x \|_2 < \rho_2 \right\}$ where $\tilde{D}_x \subseteq D_x$ such that the reduced model is exponentially stable at the origin.

According to Theorem 4, there exists ε^* such that for all $\varepsilon < \varepsilon^*$, the equilibrium point $\left[i_{Le}^T \ V_e^T \ w_e^T \ w_{qe}^T \right]^T$ of (5.48)–(5.49) with $w_{ie} \in (w_i^{min}, w_i^{max})$ and $w_{qie} \in (0, 1]$ is exponentially stable; thus completing the stability analysis of the entire DC microgrid. Note that the stability conditions (5.59), (5.62) can also provide a useful guidance for the converter or microgrid design (e.g. selection of values for C_i, L_i, R_i, etc.).

5.2.4 *Methodology for Testing the Stability Conditions*

Conditions (5.59), (5.62) might initially seem difficult to verify, mainly because they require the calculation of the equilibrium point, which, in a microgrid, is a daunting task [13, 14]. However due to the particular design of the proposed current-limiting droop controller and the boundedness properties of w_i and w_{qi}, the following pseudo-code written procedure can be used to test the conditions, for the P load case as an example (Algorithm 1):

In order to verify this methodology, the practical example that will be tested in Sect. 5.2.6 is investigated.

The system represents a DC microgrid with two boost converters in parallel feeding a P load, where each unit is equipped with the proposed controller based on the parameters specified in Table 5.3. The results of the two stability conditions are shown in Fig. 5.9, where it is clear that for any w_1 in the bounded range $\left(w_1^{min}, w_1^{max}\right)$, the expressions (5.59), (5.62) for each converter are positive, thus guaranteeing closed-loop system stability.

Test for any w_{1e} in the range (w_1^{min}, w_1^{max});
for $w_{1e} = w_1^{min} + \delta : w_1^{max} + \delta$ **do**
  Calculate w_{je} from (5.41) for all $j = 2 \ldots n$;
  if $w_j^{min} < w_{je} < w_j^{max}$ *for all* $j = 2 \ldots n$ **then**
    Calculate V_{oe} from (5.42);
    Calculate i_{Lie} from (5.44);
    Calculate $V_e = diag\{V_{ie}\}$ by combining (5.36), (5.42), (5.44) and the power P using (4.6);
    Calculate α and β from Table 5.2 and the diagonal elements of matrix D from its expression in (4.12);
    Check stability conditions;
    Verify conditions (5.59), (5.62);
  end
end

Algorithm 1: Methodology for testing stability conditions

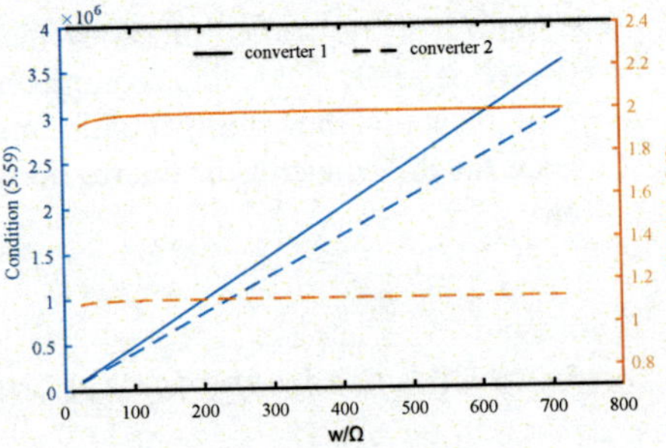

Fig. 5.9 Checking stability conditions (5.59) and (5.62)

5.2.5 *Simulation Results*

To verify the aforementioned analysis, a DC microgrid with the parameters given in Table 5.2, consisting of three boost converters feeding a common load is simulated for 20 s in MATLAB/Simulink, using the average model of the boost converter. The main tasks are to regulate the load voltage close to 400 V, share the power in a 3 : 2 : 1 ratio, and maintain an upper bound for the input current, when different types of loads (Z, I and P) are connected at the common bus.

During the first 5 s, the three converters are feeding a common Z load with a load resistance $\frac{1}{G} = 400\ \Omega$. It can be observed in Fig. 5.10a that the currents are accurately shared in a 3 : 2 : 1 ratio, having $i_1 = 0.5$ A, $i_2 = 0.33$ A, $i_3 = 0.166$ A, provided the input currents haven't reached their imposed limit yet (Fig. 5.10b). The load voltage is regulated close to the rated value as seen in Fig. 5.10c where $V_o = 399$ V.

At $t = 5$ s, the load changes to a constant I load with a load current $i_{load} = 1.5$ A. The inductor currents are still below their limit (Fig. 5.10b), and the output currents (Fig. 5.10a) keep their accurate sharing (3 : 2 : 1) with $i_1 = 0.75$ A, $i_2 = 0.5$ A, $i_3 = 0.25$ A and the load voltage remains close to 400 V, as one can observe in Fig. 5.10c where $V_o = 398.5$ V.

Table 5.2 Controller and system parameters of a DC microgrid consisting of three DC/DC boost converters feeding a common ZIP load

Parameters	Values	Parameters	Values
R_1	2.1Ω	U_1	200 V
R_2	1.9Ω	U_2	100 V
R_3	1.7Ω	U_3	240 V
L_1	2.2 mH	k_e	10
L_2	2.1 mH	m_1	0.05
L_3	2.3 mH	m_2	0.075
C_1, C_2, C_3	560 μF	m_3	0.15
i_{L1}^{max}	2 A	k_{q1}, k_{q2}, k_{q3}	1
i_{L2}^{max}	5 A	c_1, c_2, c_3	1.26×10^4
i_{L3}^{max}	2.5 A	$i_{L1}^{min}, i_{L2}^{min}, i_{L3}^{min}$	1 mA

The load changes to a constant P load at $t = 10$ s, with a load power $P = 360$ W. From Fig. 5.10c, it can be seen that the load voltage is $V_o = 399.2$ V, while the output currents (Fig. 5.10a) are $i_1 = 0.45$ A, $i_2 = 0.3$ A, $i_3 = 0.15$ A.

To test the current limitation, at $t = 15$ s, the constant P load becomes $P = 840$ W. The load voltage drops down to 397.7 V (Fig. 5.10c), and the $2:1$ power sharing is kept between converters 2 and 3, with $i_2 = 0.74$ A and $i_3 = 0.37$ A since their input currents have not reached their limit yet. However, for the first converter, the inductor current i_{L1} is successfully limited at its given upper value $i_{L1} = i_{L1}^{max} = 2$ A.

5.2.6 *Experimental Results*

A DC microgrid with the parameters given in Table 5.3, consisting of two parallel Texas Instruments DC/DC boost converters (see Appendix A) loaded by an ETPS ELP-3362F electronic load acting as a P load as shown in Fig. 5.11, is tested to experimentally evaluate the proposed control framework. A switching frequency of 60 kHz is used for the pulse-width modulation of both converters.

The main task is to achieve load voltage regulation close to the rated value $V_o^* = 48$ V and accurate power sharing among paralleled converters, while maintaining the inductor currents below their maximum values independently from the load changes. In this section, the power of the two sources satisfy the relation $P_1 = 2P_2$ and, hence, the load should be shared in a $2:1$ ratio.

To verify the effectiveness of the proposed current-limiting droop controller, it is compared to the cascaded PI approach under the same scenario. The controller parameters were calculated using the following expressions $w_{mi} = \frac{U_i}{2}\left(\frac{1}{i_{Li}^{min}} + \frac{1}{i_{Li}^{max}}\right)$ and $\Delta w_{mi} = \frac{U_i}{2}\left(\frac{1}{i_{Li}^{min}} - \frac{1}{i_{Li}^{max}}\right)$ as in [8].

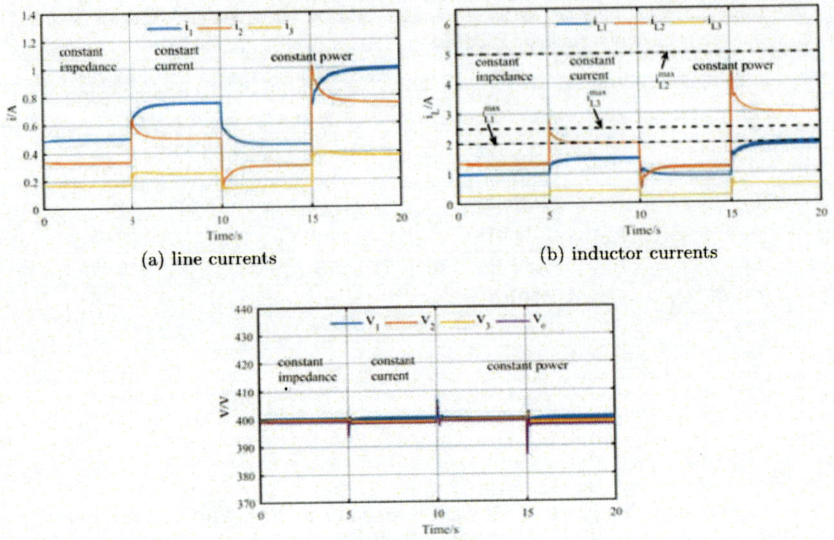

(a) line currents

(b) inductor currents

(c) output voltages

Fig. 5.10 Simulation results of the system and control states of three parallel operated DC-DC boost converters feeding a Z, I and P load

Table 5.3 Controller and experimental testbed parameters of a DC microgrid consisting of two Texas Instruments DC/DC boost converters feeding an ETPS ELP-3362F electronic load acting as a CPL

Parameters	Values	Parameters	Values
R_1	2.4 Ω	U_1	36 V
R_2	3 Ω	U_2	24 V
L_1, L_2	0.3 mH	k_{q1}, k_{q2}	1
C_1, C_2	300 μF	k_e	10
m_1	0.2	i_{L1}^{max}	1.5 A
m_2	0.4	i_{L2}^{max}	2.5 A
c_1	1012	i_{L1}^{min}	50 mA
c_2	517	i_{L2}^{min}	50 mA

Fig. 5.11 Experimental setup

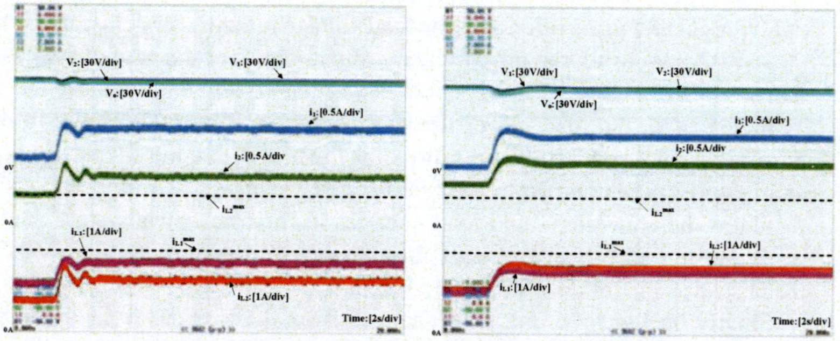

(a) dynamic response when the load demand increases from $40W$ to $60W$

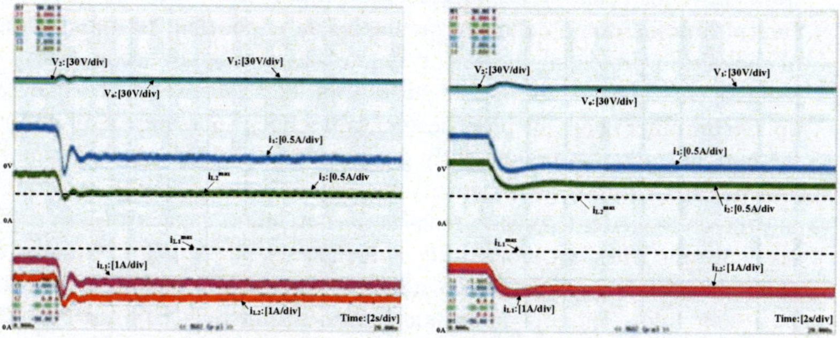

(b) dynamic response when the load demand decreases from $60W$ to $40W$

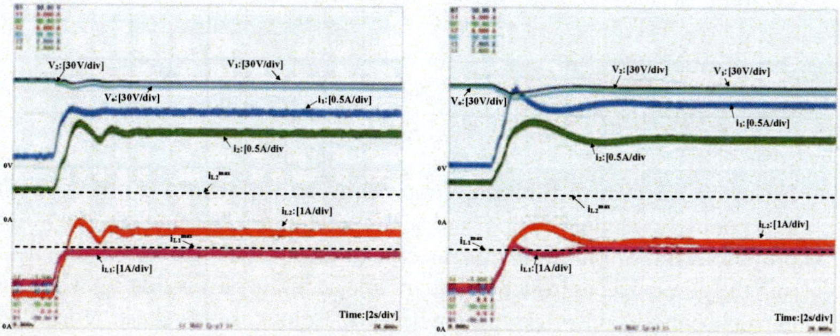

(c) dynamic response when the load demand increases from $40W$ to $80W$

Fig. 5.12 Experimental results of the system states of two parallel operated DC-DC boost converters feeding a P load using the proposed controller (left) and cascaded PI (right)

Initially, the 40 W load demand increases to 60 W and, as one can observe in Fig. 5.7a, the load is accurately shared using the proposed controller since at steady-state there is $i_1 = 2i_2$, as $i_1 \approx 0.8$ A and $i_2 \approx 0.4$ A, opposed to the case of the conventional strategy where $i_1 \neq 2i_2$. The inductor currents remain below their maximum value as imposed by the system parameters. The converters' output voltages are very tightly regulated to the reference, the load voltage remains very close to its rated value ($V_{oe} = 47.2\ V$) using the proposed controller, while for the case of the cascaded PI, it drops by 1.5 − 2 V.

In Fig. 5.7b, the load power demand decreases from 60 W down to 40 W. The power sharing is kept at the 2 : 1 ratio using the proposed controller having $i_1 \approx 0.6$ A and $i_2 \approx 0.3$ A in contrast to the cascaded PI strategy. The proposed droop control regulates the converters' and load voltages to their new steady-state values after a short transient and V_o still remains closer to the rated value, with $V_{oe} = 47.5$ V, unlike the cascaded PI framework.

Finally, in Fig. 5.7c, the load changes from 40 W to 80 W in order to test the controller performance under a large P load demand that will require a higher current from converter 1 that exceeds its technical limit. As one can observe, the proposed current-limiting droop controller maintains an upper limit for the inductor current of converter 1 protecting the device, unlike the cascaded PI droop control where the inductor current cannot be limited during transients and also leads to integrator wind-up. On the other hand, the proposed controller does not require a saturation unit and the current limitation is inherently guaranteed at all times, even during transients. The power sharing is automatically sacrificed by the proposed controller in order to protect converter 1, which reaches its maximum capacity $i_{L1} = i_{L1}^{max} = 1.5$ A. As converter 2 has not violated its own capacity $\left(i_{L2} < i_{L2}^{max}\right)$, the load demand is automatically covered and the voltage of the load is still regulated close to the rated ($V_{oe} = 46.8\ V$). This operation is achieved automatically in a decentralised way, verifying the current-limiting property and the stability analysis presented in this section.

5.3 Conclusions

In Sect. 5.1, a current-limiting droop controller for achieving power sharing among two parallel operated DC/DC boost converters in a DC microgrid application, was proposed. Based on the nonlinear dynamic model of the converters, it was proven that the proposed controller can guarantee accurate power sharing when the inductor currents of both converters remain below their maximum values. A detailed guidance for selecting the controller parameters was provided for a complete controller design. Extensive simulations were carried out and presented to validate the proposed control approach under several changes of the load demand.

In Sect. 5.2, a new current-limiting droop controller for achieving power sharing among multiple parallel operated DC-DC boost converters in a DC microgrid architecture, feeding a constant Z, I or P load, was proposed. The proposed controller additionally guarantees an inherent current limitation for each converter independently. The stability of the entire DC microgrid was analytically proven, while simulation and experimental results were presented to validate the proposed control approach under several changes of the load power demand in comparison to the conventional droop control. The superiority of the proposed current-limiting droop controller with regard to the conventional control is outlined in the following aspects:

(i) improved power sharing compared to existing strategies;
(ii) load voltage regulation closer to the rated value;
(iii) inherent current-limiting property during transients, and
(iv) proven closed-loop system stability for the nonlinear model of the DC microgrid with a Z, I or P load.

References

1. M. Mahmoodi, G.B. Gharehpetian, M. Abedi, R. Noroozian, A suitable control strategy for source converters and a novel load- generation voltage control scheme for DC voltage determination in DC distribution systems, in *2006 IEEE International Power and Energy Conference* (IEEE, New York, 2006), pp. 363–367
2. P.H. Huang, P.C. Liu, W. Xiao, M.S. El Moursi, A novel droop-based average voltage sharing control strategy for DC microgrids. IEEE Trans. Smart Grid **6**(3), 1096–1106 (2015)
3. Z. Shuai, D. He, J. Fang, Z.J. Shen, C. Tu, J. Wang, Robust droop control of DC distribution networks. IET Renew. Power Gener. **10**(6), 807–814 (2016)
4. H. Zhang, F. Mollet, C. Saudemont, B. Robyns, Experimental validation of energy storage system management strategies for a local dc distribution system of more electric aircraft. IEEE Trans. Industr. Electron. **57**(12), 3905–3916 (2010)
5. R. Ortega, A. Loria, P.J. Nicklasson, H. Sira-Ramirez, *Passivity-Based Control of Euler-Lagrange Systems: Mechanical, Electrical and Electromechanical Applications* (Springer, New York, 1998)
6. A.C. Braitor, G.C. Konstantopoulos, V. Kadirkamanathan, Power sharing of parallel operated dc-dc converters using current-limiting droop control, in *2017 25th Mediterranean Conference on Control and Automation (MED)* (2017), pp. 528–533
7. A.P.N. Tahim, D.J. Pagano, E. Lenz, V. Stramosk, Modeling and stability analysis of islanded DC microgrids under droop control. IEEE Trans. Power Electron. **30**(8), 4597–4607 (2015)
8. G.C. Konstantopoulos, Q. Zhong, Current-limiting DC/DC power converters. IEEE Trans. Control Syst. Technol. **27**(2), 855–863 (2019)
9. G.C. Konstantopoulos, Q.C. Zhong, B. Ren, M. Krstic, Bounded integral control of input-to-state practically stable nonlinear systems to guarantee closed-loop stability. IEEE Trans. Autom. Control **61**(12), 4196–4202 (2016)
10. H.K. Khalil, *Nonlinear Systems*, 3rd edn. (Pearson, London, 2014)
11. A.C. Braitor, G.C. Konstantopoulos, V. Kadirkamanathan, Current-limiting droop control design and stability analysis for paralleled boost converters in DC microgrids. IEEE Trans. Control Syst. Technol. **39**(1), 385–394 (2020)
12. A.-C. Braitor, G.C. Konstantopoulos, V. Kadirkamanathan, Admittance matrix computation and stability analysis of droop controlled dc micro-grids with CPL. IFAC-PapersOnLine **53**(2), 13525–13530 (2020)
13. M. Su, Z. Liu, Y. Sun, H. Han, X. Hou, Stability analysis and stabilization methods of DC microgrid with multiple parallel-connected DC-DC converters loaded by CPLs. IEEE Trans. Smart Grid **9**(1), 132–142 (2018)
14. N. Pogaku, M. Prodanovic, T.C. Green, Modeling, analysis and testing of autonomous operation of an inverter-based microgrid. IEEE Trans. Power Electron. **22**(2), 613–625 (2007)

Chapter 6
Stability Analysis of Parallel-Operated Bidirectional AC/DC and DC/DC Converters

In the previous chapter, the proposed controller was designed and intended for the control of unidirectional converters. Since cases have emerged where either the power converter interfaces a storage unit or interconnects two microgrids, the need for bidirectional controllers has never been more actual. The approach in Sect. 5.1.1 was to use a dynamic bounded virtual resistance, and express the maximum current i_L^{max} with respect to the input voltage U, as presented in Eq. (5.31), i.e.

$$i_L^{max} = \frac{U}{w^{min}}.$$

Since the input voltage is positive and $w \in \left(w^{min}, w^{max}\right) > 0$, the input current was always positive, the controller proving itself unsuitable for bidirectional converters. To remedy this shortcoming, in this chapter, a shift of the ellipse has been performed as one can notice in Fig. 6.1, and instead of keeping a dynamic resistance, it becomes a dynamic voltage $E \in (-E_{max}, E_{max})$ that can be both positive and negative, to ensure a bidirectional power flow.

Moreover, since in Chap. 5, the maximum current was depending on the input voltage which might vary (decrease) due to the intermittency of the energy sources availability, the converter's maximum current was prone to be limited below the technical maximum. As a result, one would not make use of the converter's full capability. In this chapter, that scenario is avoided since the current would no longer depend on the input voltage, but on the dynamic bounded virtual voltage described later on in Sect. 6.1.3.

In this chapter, a DC microgrid architecture is considered similar to Fig. 6.2, consisting of multiple energy resources in a parallel topology interfaced by both bidirectional three-phase AC/DC rectifiers and DC/DC boost converters. By considering the generic dq transformation of the AC/DC converters' dynamics and accurate non-linear model of the DC/DC converters, two novel control schemes are presented for

A.-C. Braitor, *Advanced Hierarchical Control and Stability Analysis of DC Microgrids*, Springer Theses, https://doi.org/10.1007/978-3-030-95415-4_6

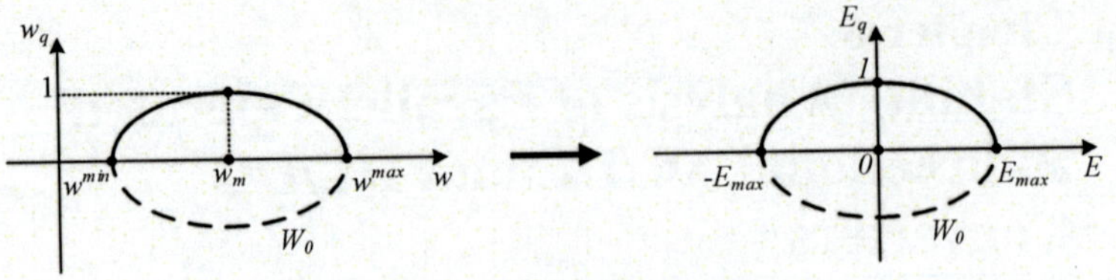

Fig. 6.1 Phase portrait shift of the bidirectional controller

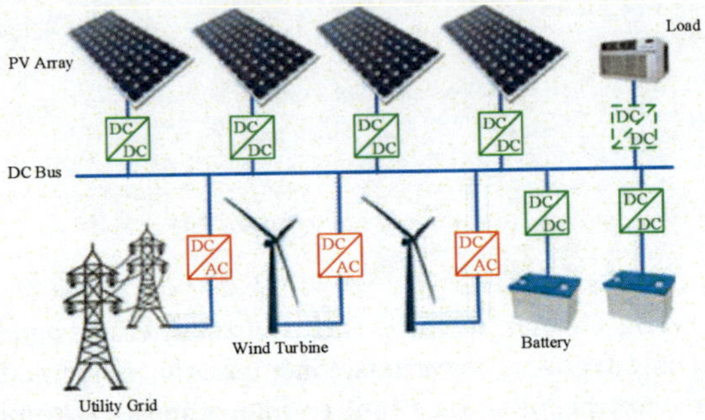

Fig. 6.2 Common configuration of a DC microgrid

each converter-interfaced unit to guarantee load voltage regulation, reactive power control, power sharing and closed-loop system stability. An inherent current-limiting capability is ensured, independently from system parameters, by means of Lyapunov methods and ultimate boundedness theory. The developed method is based on the concept of introducing a constant virtual resistance with a bounded dynamic virtual controllable voltage which can be both positive and negative to guarantee bidirectional power flow. Simulation testing is carried out in Sect. 6.1 of a DC microgrid with units feeding a common resistive load.

As CPLs are well-known to introduce instabilities into the microgrid system resulting from their negative impedance effect, in Sect. 6.2, sufficient conditions to ensure closed-loop stability with the proposed current-limiting control framework are mathematically acquired and tested in different operating scenarios. The system stability is further analysed from a graphical perspective, providing valuable insights of the CPLs influence onto the system performance and stability. The developed controller and the theoretical stability analysis are first validated in comparison with the conventional cascaded PI control technique, by simulating a three-phase AC/DC converter in parallel with a DC/DC boost converter feeding a common CPL. Experimental results are also included to further validate the benefits of the developed control strategy on a real testbed.

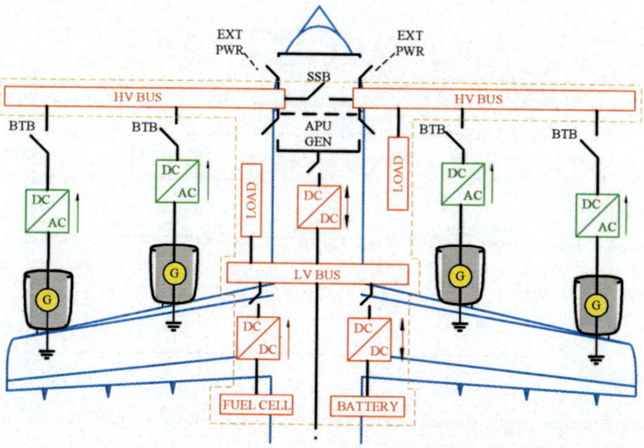

Fig. 6.3 Typical topology of an on-board DC microgrid of a hybrid electric aircraft

6.1 Bidirectional DC/DC and Three-Phase AC/DC Converters Feeding a CIL

6.1.1 *Nonlinear Model of the DC Microgrid*

Apart from the community, residential or industrial DC microgrid configuration similar to the one presented in Fig. 6.2, one can find DC microgrids in other applications such as data centres, electric vehicles and electric trains, or even electric aircrafts as shown in Fig. 6.3. The hybrid electric aircraft initiative [1], for instance, that combines conventional hydraulic and electrical systems has significantly increased, leading to recent models, such as Boeing 787 [2] and the Airbus A380 [3, 4] having more electrical components installed compared to older models.

Nevertheless, the DC microgrid under investigation is displayed in Fig. 6.4, and it consists of two main bidirectional converter units, a three-phase rectifier and a DC/DC boost converter feeding a common load. The bidirectional three-phase AC/DC rectifier consists of a boosting inductor L_s with a small parasitic resistance r_s in series for each phase, a DC output capacitor C_{rec} and six controllable switching elements that operate using PWM and capable of conducting current and power in both directions. The input voltages and currents of the rectifier are expressed as v_i and i_i, with $i = a, b, c$, while output DC voltage is denoted as V_{rec}. The bidirectional DC/DC converter has two switching elements, a parasitic resistance r_{in}, an inductor L_{bat} at the input and a capacitor C_{bat} with a line resistance R_{bat} at the output. At the input the voltage and the current of the converter are represented as U_{bat}, and i_{Lbat}, respectively, the latest being either positive or negative to allow a bidirectional power-flow.

To obtain the dynamic model of the rectifier, the average system analysis and the dq transformation can be used for three-phase voltages and currents, using Clarke and Park transformations [5]. Following [6], the dynamic model of the rectifier in

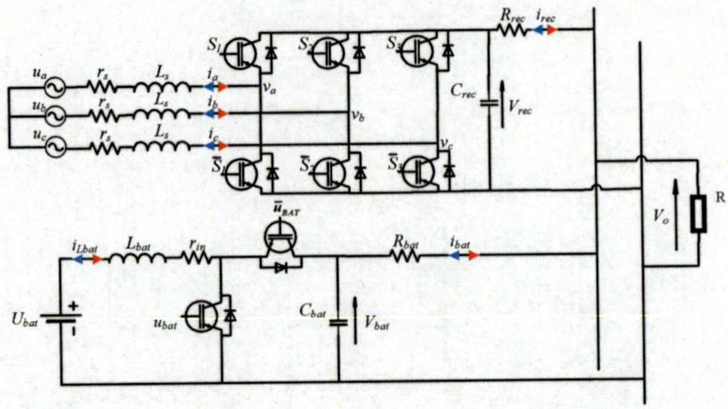

Fig. 6.4 DC microgrid under investigation consisting of a bidirectional three-phase rectifier and boost converter feeding a common load

the synchronously rotating dq frame can then be found as

$$L_s\dot{I}_d = -r_s I_d - \omega L_s I_q - m_d\frac{V_{rec}}{2} + U_d \tag{6.1}$$

$$L_s\dot{I}_q = -r_s I_q + \omega L_s I_d - m_q\frac{V_{rec}}{2} + U_q \tag{6.2}$$

$$C_{rec}\dot{V}_{rec} = \frac{3}{4}m_d I_d + \frac{3}{4}m_q I_q - i_{rec} \tag{6.3}$$

where U_d, U_q and I_d, I_q are the d and q components of the grid voltages and input currents, respectively, and m_d, m_q are the duty-ratio control inputs of the rectifier with V_d and V_q being the d and q components of the rectifier voltage $v = [v_a\ v_b\ v_c]$, respectively.

Using Kirchhoff laws and average analysis [7], the dynamic model of the bidirectional dc/dc boost converter becomes

$$L_{bat}\dot{i}_{Lbat} = U_{bat} - r_{in}i_{Lbat} - (1 - u_{bat})\,V_{bat} \tag{6.4}$$

$$C_{bat}\dot{V}_{bat} = (1 - u_{bat})\,i_{Lbat} - i_{bat} \tag{6.5}$$

It can be observed that system (6.1)–(6.5) is nonlinear, since the control inputs m_d, m_q and u_{bat} are multiplied with the system states. Assuming only the bidirectional DC/DC converter in the system, by considering a steady-state equilibrium $\left(i^e_{Lbat}, V^e_{bat}\right)$ corresponding to a duty-ratio u^e_{bat}, it results from (6.4) that $u^e_{bat} = 1 - \frac{U_{bat}}{V^e_{bat}}$, which shows that when $u_{bat} = 1$ the inductor current continuously increases, thus the system becomes unstable. Imposing a given upper bound for the inductor current is a crucial property that should be guaranteed at all times to achieve permanent device protection. A controller, equipped with the current-limiting capability while also achieving desired operation i.e. reactive power control, accurate load power sharing and tight voltage regulation, is proposed in this chapter [8].

6.1.2 Problem Description and Objectives

As mentioned earlier, a common technique to guarantee power sharing among the parallel converters, without employing communication is the droop control [9–12]. The conventional droop control method has each of the n parallel-operated power converters introducing an output voltage V_i of the form:

$$V_i = V^* - m_i \left(P_i - P_{set}\right) \tag{6.6}$$

where V^*, P_{set} are constants that represent the output reference voltage, and the set power respectively, P_i is the power drawn by the load out of each converter, m_i is the droop coefficient, with the subscript $i \in \mathcal{I}$. Nevertheless, the main concerns when employing this strategy are represented in general by the trade-off between voltage regulation and load sharing, by the influence of the system's impedance and the slow dynamic response. In addressing these problems, the droop equation in (6.6) will take the following dynamic form

$$\dot{V}_i = V^* - V_o - n_i \left(P_i - P_{set}\right) \tag{6.7}$$

where V_o is the load voltage measured at the common bus. At steady state, there is

$$m_1 P_1 = m_2 P_2 = \ldots = m_n P_n \tag{6.8}$$

which guarantees the accurate sharing of the power requested by the load. However, the technical limitations of the converters are not taken into account. Considering the power rating $P_n = P_{in}^{max}$ of a converter and the rated input voltage U_n, a limitation for the input current of each converter can be calculated. Computing such bounds for the current represents a major challenge in DC microgrids operation, since on these values depends the protection of the generating circuit or transmission system from harmful effects in cases of significant changes, such as variations in the load power demand.

6.1.3 Nonlinear Control Design and Analysis

6.1.3.1 The Proposed Controller

The purpose of the designed controller is to achieve all the aforementioned tasks without saturation units that can lead to instability. The concept behind it relies on the idea of partially decoupling the inductor current dynamics, introducing a dynamic virtual bounded controllable voltage with a constant virtual resistance for both the three-phase rectifier and the boost converter. In both cases, the dynamics of the virtual voltage will guarantee the desired upper limit for the converters' currents regardless of the direction of the power flow.

1. Three-phase bidirectional AC/DC rectifier

The control inputs m_d and m_q take the following form

$$m_d = \frac{2}{V_{rec}} \left(U_d - E_d - \omega L_s I_q + r_{vd} I_d\right) \tag{6.9}$$

$$m_q = \frac{2}{V_{rec}} \left(U_q - E_q + \omega L_s I_d + r_{vq} I_q\right) \tag{6.10}$$

where r_{vd}, r_{vq} are constant virtual resistances and E_d, E_q are virtual controllable dynamic voltages that change according to the following nonlinear dynamics:

$$\dot{E}_d = c_d g_1 (V_o, P_{REC}) E_{dq}^2 - k \left(\frac{E_d^2}{E_{maxd}^2} + w_{dq}^2 - 1 \right) E_d \tag{6.11}$$

$$\dot{E}_{dq} = -c_d g_1 (V_o, P_{REC}) \frac{E_d E_{dq}}{E_{maxd}^2} - k \left(\frac{E_d^2}{E_{maxd}^2} + E_{dq}^2 - 1 \right) E_{dq} \tag{6.12}$$

$$\dot{E}_q = c_q g_2 (Q) E_{qq}^2 - k \left(\frac{E_q^2}{E_{maxq}^2} + E_{qq}^2 - 1 \right) E_q \tag{6.13}$$

$$\dot{E}_{qq} = -c_q g_2 (Q) \frac{E_q E_{qq}}{E_{maxq}^2} - k \left(\frac{E_q^2}{E_{maxq}^2} + E_{qq}^2 - 1 \right) E_{qq} \tag{6.14}$$

with E_{dq}, E_{qq} representing additional control states and c_d, c_q, E_{maxd}, E_{maxq}, k being positive constants.

2. Bidirectional DC/DC boost converter

The control input becomes

$$u_{bat} = 1 - \frac{r_{vb} i_{bat} + U_{bat} - E_b}{V_{bat}} \tag{6.15}$$

where $r_{vb} > 0$ represents a constant virtual resistance and E_b a virtual controllable voltage which introduces the following nonlinear dynamics:

$$\dot{E}_b = cg_3\left(V_o, P_{BAT}\right)E_{bq}^2 - k\left(\frac{E_b^2}{E_{maxb}^2} + E_{bq}^2 - 1\right)E_b \tag{6.16}$$

$$\dot{E}_{bq} = -cg_3\left(V_o, P_{BAT}\right)\frac{E_b E_{bq}}{E_{maxb}^2} - k\left(\frac{E_b^2}{E_{maxb}^2} + E_{bq}^2 - 1\right)E_{bq} \tag{6.17}$$

with E_{bq} being an additional control state, c, k, E_{maxb} being positive constants and g_i, with $i = \{1, 2, 3\}$, a smooth function that describes the desired regulation scenario and has incorporated the expression of the droop control from Eq. (6.7) in the following form:

$$g_1\left(V_o, P_{REC}\right) = V^* - V_o - m_i\left(V_{rec}i_{rec} - P_{setREC}\right) \tag{6.18}$$

$$g_2\left(Q\right) = Q - Q_{set} \tag{6.19}$$

$$g_3\left(V_o, P_{BAT}\right) = V^* - V_o - m_i\left(V_{bat}i_{bat} - P_{setBAT}\right) \tag{6.20}$$

where $V_{rec}i_{rec} = P_{REC}$ and $V_{bat}i_{bat} = P_{BAT}$ represent the output power of the rectifier and bidirectional boost converter respectively.

6.1.3.2 Controller Analysis

To further understand the choice of the controller dynamics (6.16)–(6.17), consider the following Lyapunov function candidate

$$W = E_{bq}^2 + \frac{E_b^2}{E_{maxb}^2}. \tag{6.21}$$

Taking the time derivative of W and incorporating the control system (6.16)–(6.17), one obtains

$$\begin{aligned}
\dot{W} &= 2E_{bq}\dot{E}_{bq} + \frac{2E_b}{E_{maxb}^2}\dot{E}_b \\
&= -2cg_3\frac{E_b E_{bq}^2}{E_{maxb}^2} - 2k\left(\frac{E_b^2}{E_{maxb}^2} + E_{bq}^2 - 1\right)E_{bq}^2 \\
&\quad + \frac{2E_b}{E_{maxb}^2}cg_3E_{bq}^2 - 2k\frac{E_b^2}{E_{maxb}^2}\left(\frac{E_b^2}{E_{maxb}^2} + E_{bq}^2 - 1\right) \\
&= -2k\left(\frac{E_b^2}{E_{maxb}^2} + E_{bq}^2 - 1\right)\left(E_{bq}^2 + \frac{E_b^2}{E_{maxb}^2}\right).
\end{aligned} \tag{6.22}$$

From (6.22), it is clear that $\dot{W}$ is negative outside the curve

$$W_0 = \left\{E_b, E_{bq} \in \mathbb{R} : \frac{E_b^2}{E_{maxb}^2} + E_{bq}^2 = 1\right\} \tag{6.23}$$

and positive inside except from the origin, where $\dot{W} = 0$. By selecting the initial conditions E_{b0}, E_{bq0} on the curve W_0, it yields:

$$\dot{W} = 0, \Rightarrow W(t) = W(0) = 1, \forall t \geq 0, \tag{6.24}$$

which makes clear that the control states E_b and E_{bq} will start and move on the curve W_0 at all times. For convenience, the initial conditions E_{b0} and E_{bq0} will be chosen as

$$E_{b0} = 0, \;\; E_{bq0} = 1 \tag{6.25}$$

Since the control states are restricted on the curve W_0, then $E_b \in [-E_{maxb}, E_{maxb}]$ for all $t \geq 0$. The controller dynamics will result in

$$\dot{E}_b \approx c g_3 (V_o, P_{BAT}) E_{bq}^2 \tag{6.26}$$

$$\dot{E}_{bq} \approx c g_3 (V_o, P_{BAT}) \frac{E_{bq} E_b}{E_{maxb}} \tag{6.27}$$

Considering $\left(E_{b0}, E_{bq0}\right) \neq (0, 0)$, the possible equilibrium points of the controller dynamics lie on the curve W_0 that satisfy:

i. $g_3 (V_o, P_{BAT}) = 0$, that will guarantee the desired operation i.e. voltage regulation and power sharing or
ii. $\left(E_{be}, E_{bqe}\right) = (\pm E_{maxb}, 0)$ which corresponds to the case of overcurrent protection as explained in the sequel.

A similar analysis demonstrates boundedness for (6.11)–(6.14) dynamics and it will result in:

$$E_d \in [-E_{maxd}, E_{maxd}] \tag{6.28}$$

$$E_q \in \left[-E_{maxq}, E_{maxq}\right], \; \forall t \geq 0. \tag{6.29}$$

6.1.3.3 Current Limitation

1. Three-phase rectifier

For system (6.1)–(6.2), consider the following continuously differentiable function

$$V_1 = \frac{1}{2} L_s I_d^2 + \frac{1}{2} L_s I_q^2. \tag{6.30}$$

Substituting m_d, m_q from (6.9)–(6.10) into (6.1)–(6.2), and taking into account that $E_d \in [-E_{maxd}, E_{maxd}]$, $E_q \in \left[-E_{maxq}, E_{maxq}\right]$, and E_{dq}, $E_{qq} \in [0, 1]$, the time derivative of V becomes

$$\begin{aligned}\dot{V} &= -\left(r_{vd}+r_s\right)I_d^2 + E_d I_d - \left(r_{vq}+r_s\right)I_q^2 + E_q I_q \\ &\leq -\left(r_v+r_s\right)\left(I_d^2+I_q^2\right) + \begin{bmatrix} E_d & E_q \end{bmatrix}\begin{bmatrix} I_d \\ I_q \end{bmatrix} \\ &\leq -(r_v+r_s)||I||_2^2 + ||E||_2||I||_2 \end{aligned} \tag{6.31}$$

where $r_v = min\left(r_{vd}, r_{vq}\right)$, $I = \left[I_d\ I_q\right]^T$ and $E = \left[E_d\ E_q\right]^T$.

Hence

$$\dot{V}_1 < -r_s||I||_2^2,\ \forall ||I||_2 \geq \frac{||E||_2}{r_v} \tag{6.32}$$

which means that the solution of the system (6.1)–(6.2) is uniformly ultimately bounded, according to Theorem 3, with respect to the virtual voltage vector E. Since E is bounded, meaning both components E_d and E_q are bounded, then also the d and q currents, I_d and I_q remain bounded at all times.

Since $I = \left[I_d\ I_q\right]^T$, then considering the dq transformation, it results in

$$||I||_2 = \sqrt{I_d^2+I_q^2} = \sqrt{\left(\sqrt{2}I_{rms}\right)^2} = \sqrt{2}I_{rms} \tag{6.33}$$

$$||E||_2 = \sqrt{E_d^2+E_q^2} = \sqrt{\left(\sqrt{2}E_{rms}\right)^2} = \sqrt{2}E_{rms} \tag{6.34}$$

Pending a suitable choice of the control parameters, i.e. $E_{maxd} = E_{maxq} = E_{rms}^{max}$ and $r_{vd} = r_{rvq} = r_v$, then for

$$I_{rms}^{max} = \frac{E_{rms}^{max}}{r_v} \tag{6.35}$$

it is proven from the ultimate boundedness theory (6.32) that if initially the current is below the maximum allowed RMS value I_{rms}^{max}, i.e., $I_{rms}\,(0) < I_{rms}^{max}$, then

$$I_{rms}\,(t) \leq \frac{E_{rms}^{max}}{r_v} = \frac{I_{rms}^{max} r_v}{r_v} < I_{rms}^{max},\ \forall t > 0. \tag{6.36}$$

Hence, the input current of the rectifier is always limited below I_{rms}^{max} with the appropriate choice of E_{rms}^{max} and r_v given in (6.35), ensuring protection at all times. By maintaining an upper limit for E_d and E_q from the proposed dynamics (6.11)–(6.14), both the closed-loop system stability and the desired current-limiting property are achieved. Since the dynamics (6.11)–(6.14) are analysed using Lyapunov theory that induces invariant sets, the required bounds for E_d and E_q are guaranteed without applying additional saturation units. In addition, the proposed controller slows down the integration near the limits, and therefore, it does not suffer from integrator windup issues, which may lead to instability. This is a crucial property that distinguishes the proposed controller with traditional current-limiting approaches that incorporate current saturation units.

2. Bidirectional boost converter
By applying the proposed controller expression (6.15) into the bidirectional converters dynamics (6.4), the closed-loop system equation for the inductor current i_{Lbat} takes the following form

$$L_{bat}\dot{i}_{Lbat} = -(r_{vb} + r_s)\, i_{Lbat} + E_b, \tag{6.37}$$

and it becomes clear that r_{vb} represents a constant virtual resistance in series with the converter inductor L_{bat}.

To investigate how the selection of the virtual resistance and the bounded controller dynamics of E_b are related to the desired overcurrent protection, let the following Lyapunov function candidate

$$V_2 = \frac{1}{2} L_{bat} i_{Lbat}^2 \tag{6.38}$$

for closed-loop current dynamics (6.37). The time derivative of V_2 yields

$$\begin{aligned}\dot{V}_2 &= L_{bat} i_{Lbat} \dot{i}_{Lbat} = -(r_{vb} + r_s)\, i_{Lbat}^2 + E_b i_{Lbat} \\ &\leq -(r_{vb} + r_s)\, i_{Lbat}^2 + |E_b||i_{Lbat}| \leq -(r_{vb} + r_s)\, i_{Lbat}^2 + E_{maxb}|i_{Lbat}|,\end{aligned} \tag{6.39}$$

given the bounded $E_b \in [-E_{maxb}, E_{maxb}]$, which implies that

$$\dot{V}_2 < -r_s i_{Lbat}^2, \quad \forall |i_{Lbat}| > \frac{E_{maxb}}{r_{vb}}. \tag{6.40}$$

By virtue of Theorem 3 and according to (6.40), the solution $i_{Lbat}(t)$ is uniformly ultimately bounded. So, if initially $|i_{Lbat}(0)| \leq \frac{E_{maxb}}{r_{vb}}$, then it holds that

$$|i_{Lbat}(t)| \leq \frac{E_{maxb}}{r_{vb}}, \quad \forall t > 0, \tag{6.41}$$

because of the invariant set property. Based on the desired overcurrent protection, it should hold true that

$$|i_{Lbat}(t)| \leq i_{Lbat}^{max}, \quad \forall t > 0, \tag{6.42}$$

for a given maximum value i_{Lbat}^{max} of the inductor current. By substituting (6.41) into (6.42), one can clearly select the parameters E_{maxb} and r_{vb} in the proposed controller in order to satisfy

$$E_{maxb} = r_{vb} i_{Lbat}^{max}. \tag{6.43}$$

Any selection of the constant and positive parameters E_{maxb} and r_{vb} that satisfy (6.43) results in the desired overcurrent protection (6.42) of the converter's inductor current regardless the load magnitude or system parameters.

From the closed-loop dynamics (6.37) combined with (6.16)–(6.17) at steady state, there is $g_3(V_o, P_{bat}) = 0$, then $E_b = E_{be}$ on the curve W_0 and the value of the inductor current becomes $i_{Lbat}^e = \frac{E_{be}}{r_{vb}}$. But since $E_{be} \in [-E_{maxb}, E_{maxb}]$, then the inductor current can be both positive and negative, thus, ensuring the two-way operation of the bidirectional converter. When $E_{be} = -E_{maxb}$, then $i_{Lbat}^e = -\frac{E_{maxb}}{r_{vb}} = -i_{Lbat}^{max}$ that corresponds to the overcurrent protection in both directions of the current.

Compared to existing conventional overcurrent protection control strategies, it has been mathematically proven according to the ultimate boundedness theory that the proposed controller maintains the current limited during transients and does not require limiters or saturation units which are prone to yield instability in the system, thus highlighting the superiority of the proposed control design.

6.1.4 Simulation Results

To test the proposed controller, by considering the average of the bidirectional DC/DC boost converter and the three-phase AC/DC rectifier, the DC microgrid displayed in Fig. 6.4 is studied having the parameters specified in Table 6.1, with the control parameters I_{max} and r_v chosen as specified in Sect. 6.1.3. The aim is to achieve tight voltage regulation around the reference value $V^* = 400$ V, accurate power sharing in a $2:1$ ratio among the paralleled AC/DC and DC/DC converters at the load bus while also assuring protection against overcurrents. The model has been implemented in Matlab/Simulink and simulated for 30 s, considering a full testing scenario.

During the first 5 s, it can be observed in Fig. 6.5b that the load voltage V_o is kept close the reference value of 400 V. The power sharing is accurately guaranteed (Fig. 6.5c) in a $2:1$ manner having $i_{BAT} \approx 0.34$ A and $i_{REC} \approx 0.17$ A, since the input currents haven't reached their imposed limits yet as shown in Fig. 6.5a.

For the next 10 s the operation principle of the battery is simulated. The direction of the power flow is reversed to allow the battery to charge and discharge. At $t = 5$ s the power set by the battery controller becomes negative $P_{setBAT} = -500$ W, thus leaving the battery to be supplied by the three-phase rectifier.

The input current goes to the negative side, while the rectifier's input current increases to satisfy the new amount of power requested in the network (Fig. 6.5a). The power sharing ratio between the battery and the rectifier disappears since the current of the battery changes its direction, and becomes negative as shown in Fig. 6.5b. The load voltage remains closely regulated to the desired 400 V value. After 5 s the set

Table 6.1 Controller and system parameters of a DC microgrid consisting of a bidirectional DC/DC boost and a three-phase AC/DC converter feeding a common CIL

Parameters	Values	Parameters	Values
U_{RMS}	110 V	U_{bat}	200 V
R_{phase}	0.5 Ω	$R_{bat,rec}$	1.1 Ω
L_{phase}	2.2 mH	L_{bat}	2.3 mH
I_{max}^{RMS}	3.3 A	I_{max}^{bat}	5 A
C_{rec}	300 μF	C_{bat}	500 μF
n_{rec}	0.015	n_{bat}	0.0075
P_{load}	200 W	k	1000
c_d	100000	c_{bat}	100
c_q	5000	r_{vbat}	5 Ω

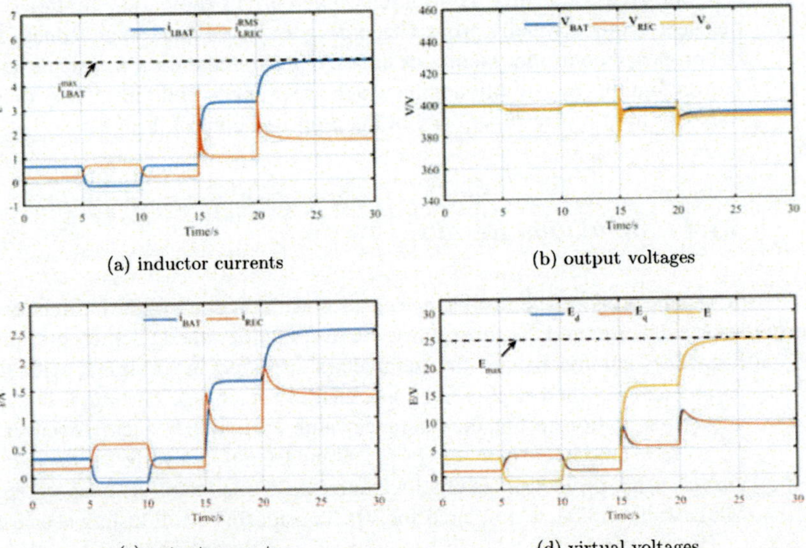

(a) inductor currents

(b) output voltages

(c) output currents

(d) virtual voltages

Fig. 6.5 Simulation results of the DC microgrid system

value of the power returns to its initial 0 W value, allowing the battery to return to its former discharging state. The power sharing ratio comes back to 2 : 1 as displayed in Fig. 6.5b.

At $t = 15$ s the power requested by the load increases $P_{load} = 1000$ W and, thus, more power is needed from the battery and the three-phase rectifier to be injected in the microgrid. The load voltage drops down to 395 V according to Fig. 6.5b, while the input currents increase and, therefore, the power injected increases at the common bus (Fig. 6.5a) but keeps the imposed sharing between the two sources, the battery and rectifier, to the desired proportion of 2 : 1 having $i_{BAT} \approx 1.65\,\text{A}$ and $i_{REC} \approx 0.82\,\text{A}$, as presented in Fig. 6.5c given the fact that none of the inductor currents have reached their maximum allowed current.

To test the input current protection capability, the power demanded by the load is further increased. Thus, at $t = 20\,\text{s}$ the power requested by the load reaches a higher value than before, $P_{load} = 1600\,\text{W}$, forcing the battery and the three-phase rectifier to increase their power injection at the load bus. As noticed in Fig. 6.5a, the input current of the battery reaches its limit $i_{LBAT} = i_{LBAT}^{max} = 5\,\text{A}$, and the power sharing is sacrificed (Fig. 6.5c) to ensure uninterruptible power supply to the load. The load voltage remains within the desired range, $V_o = 391\,\text{V}$ with a voltage drop of 9 V, which is about 2%.

Consequently, to further verify the theory presented, the controller states E_d, E_q, and E are presented in Fig. 6.5d. When the input current of the battery reaches its maximum, the virtual voltage of battery also arrives at its imposed limit $E_{BAT} = E_{maxBAT} = i_{LBAT}^{max} r_{vBAT} = 25\,\text{V}$.

6.2 Bidirectional DC/DC and Three-Phase AC/DC Converters Feeding a CPL

6.2.1 Dynamic Model

The DC microgrid shown in Fig. 6.2 is extended from two parallel-operated converters to any finite number of parallel-operated converters. The configuration of the DC microgrid under investigation is shown in Fig. 6.6, containing n bidirectional three-phase rectifiers and m bidirectional DC/DC boost converters feeding a constant power load, where L_{si} is the inductor at the input, a DC output capacitor C_i with a line resistance R_i and six controllable switching elements that operate using PWM and capable of conducting current and power in both directions. The input voltages and currents of the rectifier are expressed as v_{ai}, v_{bi}, v_{ci} and i_{ai}, i_{bi}, i_{ci}, while output dc voltage is denoted as V_i with $i \in \{1, 2, ..., n\}$. The bidirectional DC/DC converters have two switching elements, an inductor L_j at the input and a capacitor C_j with a line resistance R_j at the output, while V_j is the output voltage, where $j \in \{n+1, n+2, \ldots, n+m\}$. At the input, the voltage and the current of the converter are represented as U_j, and i_{Lj}, respectively, with the latter being either positive or negative to allow a bidirectional power-flow.

To obtain the dynamic model of the rectifier, the average system analysis and the dq transformation can be used for three-phase voltages and currents, using Clarke and Park transformations [5]. Following [6], the mathematical model of the rectifiers in the dq coordinates is set up, in matrix form as

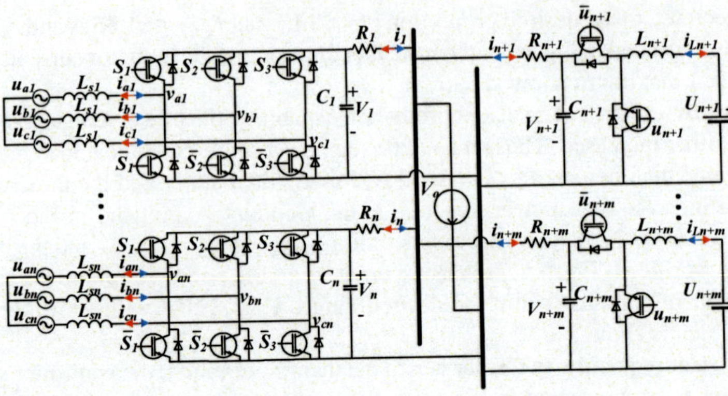

Fig. 6.6 Parallel operated three-phase AC/DC and bidirectional DC/DC boost converters feeding a common constant power load

$$L_s\dot{I}_d = -\omega L_s I_q - \frac{1}{2}m_d V_r + U_d \tag{6.44}$$

$$L_s\dot{I}_q = \omega L_s I_d - \frac{1}{2}m_q V_r \tag{6.45}$$

$$C\dot{V}_r = \frac{3}{4}m_d I_d + \frac{3}{4}m_q I_q - i_r \tag{6.46}$$

where $i_r = [i_1 \ldots i_n]^T$, $V_r = [V_1 \ldots V_n]^T$, $L_s = diag\{L_{si}\}$, $C_r = diag\{C_i\}$, $\omega = diag\{\omega_i\}$ is the rotating speed, $U_d = [U_{d1} \ldots U_{dn}]^T$ is the amplitude of the three-phase AC voltage source when voltage orientation on the d axis is considered and $I_d = [I_{d1} \ldots I_{dn}]^T$, $I_q = [I_{q1} \ldots I_{qn}]^T$ are the d and q components of the AC source currents, respectively, and $m_d = diag\{m_{di}\}, m_q = diag\{m_{qi}\}$ are the duty-ratio control inputs of the rectifier with V_d and V_q being the d and q components of the rectifier voltage $v = [v_a \; v_b \; v_c]$, respectively.

Using Kirchhoff laws and average analysis [7], the dynamic model, in matrix form, of the bidirectional DC/DC boost converter becomes

$$L\dot{i}_L = U_b - (I_m - u)\,V_b \tag{6.47}$$

$$C_b\dot{V}_b = (I_m - u)\,i_L - i_b \tag{6.48}$$

where $i_L = [i_{L(n+1)} \ldots i_{L(n+m)}]^T$, $V_b = [V_{n+1} \ldots V_{n+m}]^T$, $i_b = [i_{n+1} \ldots i_{n+m}]^T$, $U_b = [U_{n+1} \ldots U_{n+m}]^T$, $L = diag\left\{L_j\right\}, C_b = diag\left\{C_j\right\}, u = diag\left\{u_j\right\}$. One can observe that system (6.44)–(6.46), (6.47)–(6.48) is nonlinear, since the control inputs m_{di}, m_{qi} and u_j are multiplied with the system states, $\left(I_d, I_q, V_r\right)$, and (i_L, V_b) respectively.

As the AC/DC and DC/DC converters supply a CPL, the power balance equation becomes

$$P = V_o \sum_{k=1}^{n+m} i_k, \tag{6.49}$$

$$i_k = \frac{V_k - V_o}{R_k} \tag{6.50}$$

where V_k, i_k represent the output voltages and currents, respectively, with $k \in \{1, 2, \ldots, n+m\}$, V_o is the load voltage, and P is constant and represents the power of the CPL. Consider now the following assumptions:

Assumption 6 It holds that

$$\left(\sum_{k=1}^{n+m} \frac{V_k}{R_k} \right)^2 > 4P \sum_{k=1}^{n+m} \frac{1}{R_k}. \tag{6.51}$$

Thus, substituting the output current i_k from (6.50) into (6.49), one can obtain the following expression for the load voltage given by the real solutions of the second order polynomial

$$V_o = \frac{\sum_{k=1}^{n+m} \frac{V_k}{R_k} \pm \sqrt{\left(\sum_{k=1}^{n+m} \frac{V_k}{R_k} \right)^2 - 4P \sum_{k=1}^{n+m} \frac{1}{R_k}}}{2 \sum_{k=1}^{n+m} \frac{1}{R_k}} \tag{6.52}$$

Assumption 7 Let $I_k^{max} = \{I_{rms1}^{max}, \ldots, I_{rmsn}^{max}, i_{L(n+1)}^{max}, \ldots, i_{L(n+m)}^{max}\}$ be the maximum input current of each converter (maximum RMS current for AC/DC converters and maximum inductor current for DC/DC converters). Since for three-phase rectifiers $V_i \geq 2U_{di}$ and for boost converters $V_j \geq U_j$, let

$$min\{2U_{di}, U_j\} - I_k^{max} R_k > \frac{\sum_{k=1}^{n+m} \frac{V_k}{R_k} \pm \sqrt{\left(\sum_{k=1}^{n+m} \frac{V_k}{R_k} \right)^2 - 4P \sum_{k=1}^{n+m} \frac{1}{R_k}}}{2 \sum_{k=1}^{n+m} \frac{1}{R_k}} \tag{6.53}$$

hold, for every $k \in \{1, 2, \ldots, n+m\}$.

The load voltage has two solutions, a high voltage and a low voltage, with the high voltage representing the feasible solution because of Assumption 7, which gives $V_o \geq min\{2U_{di}, U_j\} - I_k^{max} R_k$. Therefore, the voltage of the load can be described as

$$V_o = \frac{\sum_{k=1}^{n+m} \frac{V_k}{R_k} + \sqrt{\left(\sum_{k=1}^{n+m} \frac{V_k}{R_k} \right)^2 - 4P \sum_{k=1}^{n+m} \frac{1}{R_k}}}{2 \sum_{k=1}^{n+m} \frac{1}{R_k}} \tag{6.54}$$

Considering an equilibrium point $\left(I_{die}, I_{qie}, i_{Lje}, V_{ie}, V_{je}\right)$ for constant control inputs m_{di}, m_{qi}, u_j, by taking the partial derivative of the output current i_k from (6.50) with respect to the capacitor voltage V_k, we obtain the admittance matrix [13, 14] as in (4.11):

$$Y = R^{-1}\left(I_{n+m} - \mathbf{1}_{(n+m)\times(n+m)}D\right) \tag{6.55}$$

with matrix D for the CPL case following from Eq. (4.12) as

$$D = \frac{1}{2\sum_{k=1}^{n+m}\frac{1}{R_k}}\left(R^{-1} + \frac{\sum_{k=1}^{n+m}\frac{V_k}{R_k}}{\sqrt{\left(\sum_{k=1}^{n+m}\frac{V_{ke}}{R_k}\right)^2 - 4P\sum_{k=1}^{n+m}\frac{1}{R_k}}}R^{-1}\right) \tag{6.56}$$

where $\sqrt{\left(\sum_{k=1}^{n+m}\frac{V_k}{R_k}\right)^2 - 4P\sum_{k=1}^{n+m}\frac{1}{R_k}} > 0$ according to Assumption 6. Since R is a diagonal positive-definite matrix, then it is clear that matrix D is a positive-definite diagonal matrix, with eigenvalues of the form

$$\lambda_{Dk} = \frac{1}{2\sum_{k=1}^{n+m}\frac{1}{R_k}}\left(\frac{1}{R_k} + \frac{\sum_{k=1}^{n+m}\frac{V_k}{R_k}}{\sqrt{\left(\sum_{k=1}^{n+m}\frac{V_{ke}}{R_k}\right)^2 - 4P\sum_{k=1}^{n+m}\frac{1}{R_k}}}\frac{1}{R_k}\right), \tag{6.57}$$

$\forall k = 1, \ldots, n+m.$

6.2.2 *Nonlinear Control Design and Analysis*

The end goal of the designed controller is to achieve accurate distribution of the load power and tight load voltage regulation close to the rated value, ensuring that the current of each converter does not violate certain bounds. This concept is based on the idea of partially decoupling the inductor current dynamics, introducing a constant virtual resistance with a bounded controllable voltage for both the bidirectional three-phase AC/DC and the DC/DC boost converters. In both cases, the dynamics of the controllable virtual voltage will guarantee the desired upper bound for the converters' currents regardless of the direction of the power flow.

6.2.2.1 Three-Phase Rectifier

Although a current-limiting controller was recently proposed in [5], it only allows unidirectional power flow, which is a significant limitation when storage units are introduced or the AC/DC converter represents an interface between a DC and an AC microgrid. To overcome this problem, here the control inputs m_{di} and m_{qi}, with $i \in \{1, 2, ..., n\}$ are proposed to take the following form

$$m_{di} = \frac{2}{V_i}\left(U_{di} - E_{di} - \omega_i L_{si} I_{qi} + r_{vi} I_{di}\right) \tag{6.58}$$

$$m_{qi} = \frac{2}{V_i}\left(\omega_i L_{si} I_{di} + r_{vi} I_{qi}\right) \tag{6.59}$$

where $r_{vi} > 0$ is a constant virtual resistance and E_{di} a virtual voltage that change according to the following nonlinear dynamics:

$$\dot{E}_{di} = c_{di}\left(V^{*} - V_o - d_i\left(\frac{3}{2}\frac{U_{di} E_{di}}{r_{vi}} - P_{seti}\right)\right)E_{dqi}^2 \tag{6.60}$$

$$\dot{E}_{dqi} = -c_{di}\left(V^{*} - V_o - d_i\left(\frac{3}{2}\frac{U_{di} E_{di}}{r_{vi}} - P_{seti}\right)\right)\frac{E_{di} E_{dqi}}{E_{maxi}^2} - k_i c_{di}\left(\frac{E_{di}^2}{E_{maxi}^2} + E_{dqi}^2 - 1\right)E_{dqi} \tag{6.61}$$

with E_{dqi} representing an additional control state, V^{*} the load voltage reference, P_{seti} the set output power, d_i the droop coefficient, and c_{di}, E_{maxi}, k_i being positive constants. The proposed controller introduces the desired droop expression via the input m_{di}, while it forces the current I_{qi} to zero through m_{qi} in order to guarantee unity power factor operation, since $Q_i = \frac{3}{2} U_{di} I_{qi}$.

6.2.2.2 Bidirectional DC/DC Boost Converter

Following a similar control framework with the AC/DC converter, for the DC/DC boost converter the control input u_j, with $j \in \{n+1, ..., n+m\}$, becomes

$$u_j = 1 - \frac{r_{vj} i_{Lj} + U_j - E_j}{V_j} \tag{6.62}$$

where $r_{vbj} > 0$ represents a constant virtual resistance and E_j a virtual controllable voltage

$$\dot{E}_j = c_j\left(V^{*} - V_o - d_j\left(\frac{U_j E_j}{r_{vbj}} - P_{setj}\right)\right)E_{bqj}^2 \tag{6.63}$$

$$\dot{E}_{bqj} = -c_j\left(V^{*} - V_o - d_j\left(\frac{U_j E_j}{r_{vbj}} - P_{setj}\right)\right)\frac{E_j E_{bqj}}{E_{maxj}^2} - k_j c_j\left(\frac{E_j^2}{E_{maxj}^2} + E_{bqj}^2 - 1\right)E_{bqj} \tag{6.64}$$

where E_{bqj} being an additional control state, P_{setj} the set output power, d_j the droop coefficient, and c_j, k_j, E_{maxj} positive constants. Compared to the robust droop controller [12], the proposed strategy does not require the measurement of the output current i_i, i_j of each converter, thus leading to a simpler implementation. It is highlighted that a second controller state E_{dq}, E_{bq} is based on the bounded integral controller concept [15]. For more details on the bounded dynamics of the control states the reader is referred to [15] where it is shown that the control states are guaranteed to

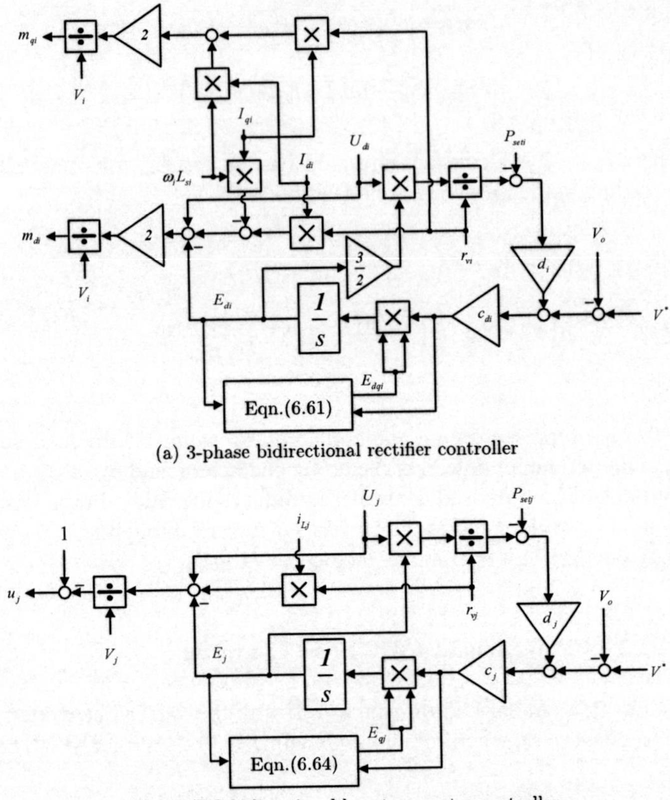

(a) 3-phase bidirectional rectifier controller

(b) DC/DC bidirectional boost converter controller

Fig. 6.7 Block diagrams with the control implementation of the controllers

stay within their imposed bounds $E_{di} \in [-E_{maxi}, E_{maxi}]$, $E_j \in \left[-E_{maxj}, E_{maxj}\right]$ and $E_{dqi}, E_{bqj} \in [0, 1]$ for all $t \geq 0$, given typical initial conditions $E_{di} = E_j = 0$ and $E_{dqi} = E_{bqj} = 1$. The block diagram depicting the controller implementation, measurement and actuation parts is presented in Fig. 6.7. Having introduced the proposed control schemes, consider the additional assumptions for the system:

Assumption 8 For every $E_{die} \in (-E_{maxi}, E_{maxi})$ and $E_{je} \in (-E_{maxj}, E_{maxj})$ constant, satisfying

$$d_{d1}\frac{3}{2}\left(\frac{U_{d1}E_{d1e}}{r_{vd1}} - P_{set1}\right) = \ldots = d_{dn}\frac{3}{2}\left(\frac{U_{dn}E_{dne}}{r_{vdn}} - P_{setn}\right) =$$
$$d_{n+1}\left(\frac{U_{n+1}E_{(n+1)e}}{r_{v(n+1)}} - P_{set(n+1)}\right) = \ldots = d_{n+m}\left(\frac{U_{n+m}E_{(n+m)e}}{r_{v(n+m)}} - P_{set(n+m)}\right)$$

there exists a unique steady-state equilibrium point, denoted

$$\left(I_{die}, I_{qie}, i_{Lje}, V_{ie}, V_{je}, E_{die}, E_{dqie}, E_{je}, E_{bqje}\right)$$

corresponding to a load voltage regulation V_{oe} from (6.54), where $E_{dqie}, E_{bqje} \in (0, 1], \forall k = 1, \ldots, n+m$.

Assumption 9 For $\forall k = 1 \ldots n+m$ it holds that $\frac{U_k d_k}{\alpha_k R_k} > 1$, with $\alpha_k = 3$ when $k = 1 \ldots n$ and $\alpha_k = 1$ when $k = n+1 \ldots n+m$.

For the selection of E_{maxi} and E_{maxj} the following condition should hold

$$|E_{maxk}| < U_k, \forall k = 1 \ldots n+m. \tag{6.65}$$

Remark 6 Similar to Chap. 5, one might notice from the previous section that the desired current-limiting feature was proven by employing the ultimate boundedness theory and making use of the parasitic resistances of each converter. Although in this section they have been ignored, one can easily guarantee the current-limiting property of each converter by simply rewriting the constant virtual resistance r_v, as $r_v = \overline{r_v} + \epsilon$, for an arbitrarily small positive constant ϵ.

6.2.3 *Stability Analysis*

By applying the proposed controller (6.58)–(6.61), (6.62)–(6.64) into the DC micro-grid dynamics (6.44)–(6.46), (6.47)–(6.48) the closed-loop system can be written in the following matrix form

$$\begin{bmatrix} \dot{I}_d \\ \dot{I}_q \\ \dot{i}_L \\ \dot{V}_r \\ \dot{V}_b \end{bmatrix} = \begin{bmatrix} L_s^{-1}\left(-r_{vd} I_d + E_d\right) \\ -L_s^{-1} r_{vd} I_q \\ L_b^{-1}\left(-r_{vb} i_L + E_b\right) \\ \frac{3}{2} C_r^{-1}\left[V_r\right]^{-1}\left(\left(\left[U_d\right]-\left[E_d\right]+r_{vd}\left[I_d\right]\right) I_d - r_{vd} I_q^2\right) - C_r^{-1} i_r \\ C_b^{-1}\left[V_b\right]^{-1}\left(r_{vb}\left[i_L\right]+\left[U_b\right]-\left[E_b\right]\right) i_L - C_b^{-1} i_b \end{bmatrix} \tag{6.66}$$

$$\begin{bmatrix} \dot{E}_d \\ \dot{E}_{dq} \\ \dot{E}_b \\ \dot{E}_{bq} \end{bmatrix} = \begin{bmatrix} c_d\left[E_{dq}\right]^2\left(V^*\mathbf{1}_n - V_o\mathbf{1}_n - d_d\left(\frac{3}{2} r_{vd}^{-1}\left[U_d\right] E_d - P_{setd}\right)\right) \\ -c_d E_{maxd}^{-2}\left[E_d\right]\left[E_{dq}\right]\left(V^*\mathbf{1}_n - V_o\mathbf{1}_n - d_d\left(\frac{3}{2} r_{vd}^{-1}\left[U_d\right] E_d - P_{setd}\right)\right) \\ c_b\left[E_{bq}\right]^2\left(V^*\mathbf{1}_m - V_o\mathbf{1}_m - d_b\left(r_{vb}^{-1}\left[U_b\right] E_b - P_{setb}\right)\right) \\ -c_b E_{maxb}^{-2}\left[E_b\right]\left[E_{bq}\right]\left(V^*\mathbf{1}_m - V_o\mathbf{1}_m - d_b\left(r_{vb}^{-1}\left[U_b\right] E_b - P_{setb}\right)\right) \end{bmatrix} - \begin{bmatrix} \mathbf{0}_n \\ k_d c_d\left(E_{maxd}^{-2}\left[E_d\right]^2+\left[E_{dq}\right]^2 - I_n\right) E_{dq} \\ \mathbf{0}_n \\ k_b c_b\left(E_{maxb}^{-2}\left[E_b\right]^2+\left[E_{bq}\right]^2 - I_m\right) E_{bq} \end{bmatrix} \tag{6.67}$$

with $d_d = diag\{d_i\}$, $d_b = diag\{d_j\}$, $k_d = diag\{k_i\}$, $k_b = diag\{k_j\}$, $E_d = [E_{d1} \dots E_{dn}]^T$, $E_{dq} = \left[E_{dq1} \dots E_{dqn}\right]^T$, $E_{bq} = \left[E_{bq(n+1)} \dots E_{bq(n+m)}\right]^T$, $r_{vd} = diag\{r_{vi}\}$, $c_d = diag\{c_{di}\}$, $c_b = diag\{c_j\}$, $E_{maxd} = diag\{E_{maxi}\}$, $E_{maxb} = diag\{E_{maxj}\}$, $P_{setd} = [P_{set1} \dots P_{setn}]^T$, $P_{setb} = \left[P_{set(n+1)} \dots P_{set(n+m)}\right]^T$.

Consider an equilibrium point $\left[I_{de}^T \, I_{qe}^T \, i_{Le}^T \, V_{re}^T \, V_{be}^T \, E_{de}^T \, E_{dqe}^T \, E_{be}^T \, E_{bqe}^T\right]$ calculated from (6.66)–(6.67) at the steady-state, satisfying Assumption 8. By setting $\varepsilon = \frac{1}{min\{c_k\}}$, there exists $\delta_d = diag\{\delta_i\} \geq 0$ and $\delta = diag\left\{\delta_j\right\} \geq 0$ such that $c_d = \frac{1}{\varepsilon} I_n + \delta_d$ and $c = \frac{1}{\varepsilon} I_m + \delta$. Thus (6.67) becomes

$$
\begin{bmatrix} \varepsilon \dot{E} \\ \varepsilon \dot{E}_q \end{bmatrix} = \begin{bmatrix} I_{n+m} + \bar{\delta} & \mathbf{O}_{(n+m)\times(n+m)} \\ \mathbf{O}_{(n+m)\times(n+m)} & I_{n+m} + \bar{\delta} \end{bmatrix} \times \left(\begin{bmatrix} \left[E_q\right]^2 \left((V^* - V_o)\,\mathbf{1}_{n+m} - d\left(H r_v^{-1}[U]\,E - P_{set}\right)\right) \\ -E_{max}^{-2}[E]\left[E_q\right]\left((V^* - V_o)\,\mathbf{1}_{n+m} - d\left(H r_v^{-1}[U]\,E - P_{set}\right)\right) \end{bmatrix} - \begin{bmatrix} \mathbf{0}_n \\ k\left(E_{max}^{-2}[E]^2 + \left[E_q\right]^2 - I_m\right) E_q \end{bmatrix} \right) \tag{6.68}
$$

where $H = \begin{bmatrix} \frac{3}{2} I_n & \mathbf{0}_{n\times m} \\ \mathbf{0}_{m\times n} & I_m \end{bmatrix}$, $\bar{\delta} = diag\{\delta_d, \delta\}$, $d = diag\{d_k\}$, $k = diag\{k_k\}$, $E = \left[E_d^T \, E_b^T\right]^T$, $E_q = \left[E_{dq}^T E_{bq}^T\right]^T$, $r_v = diag\{r_{vk}\}$, $U = \left[U_d^T U_b^T\right]^T$, $E_{max} = diag\{E_{maxk}\}$, $P_{set} = \left[P_{setd}^T \, P_{setb}^T\right]^T$.

Hence, the closed-loop system equations and can be written as

$$
\dot{x} = f(x, z) \tag{6.69}
$$

$$
\varepsilon \dot{z} = g(x, z) \tag{6.70}
$$

where $x = \begin{bmatrix} I_d - I_{de} \\ I_q \\ i_L - i_{Le} \\ V_r - V_{re} \\ V_b - V_{be} \end{bmatrix}$ and $z = \begin{bmatrix} E_d - E_{de} \\ E_{dq} - E_{dqe} \\ E - E_e \\ E_{bq} - E_{bqe} \end{bmatrix}$. For arbitrarily large values of the controller gains c, c_d the value of ε is small and therefore (6.69)–(6.70) can be investigated as a singularly perturbed system using two-time-scale analysis [16].

Considering f, g being continuously differentiable in the domain $(x, z, \varepsilon) \in D_x \times D_z \times [0, \varepsilon_0]$, when the controller gains c, c_d is selected sufficiently large, then $\varepsilon \to 0$ and, based on singular perturbation theory, g will have an algebraic form of $0 = g(x, z)$ as follows

$$\begin{bmatrix} \mathbf{0}_{n+m} \\ \mathbf{0}_{n+m} \end{bmatrix} = \begin{bmatrix} [E_q]^2 \left((V^* - V_o)\,\mathbf{1}_{n+m} - d\left(Hr_v^{-1}[U]\,E - P_{set}\right)\right) \\ -E_{max}^{-2}[E]\left[E_q\right]\left((V^* - V_o)\,\mathbf{1}_{n+m} - d\left(Hr_v^{-1}[U]\,E - P_{set}\right)\right) \end{bmatrix} - \begin{bmatrix} \mathbf{0}_n \\ k\left(E_{max}^{-2}[E]^2 + \left[E_q\right]^2 - I_m\right)E_q \end{bmatrix} \tag{6.71}$$

The roots of the above system can be computed as shown below

$$\begin{bmatrix} \overline{E} \\ \overline{E}_q \end{bmatrix} = \begin{bmatrix} H^{-1}r_v\,[U]^{-1}\left(d^{-1}(V^* - V_o)\,\mathbf{1}_{n+m} + P_{set}\right) \\ \left(I_{n+m} - E_{max}^{-2}\left[\overline{E}\right]^2\right)^{-\frac{1}{2}}\mathbf{1}_{n+m} \end{bmatrix} \tag{6.72}$$

These roots can also be written as $z = h(x)$ with $\overline{E}_{die} \in (-E_{maxi}, E_{maxi})$, $\overline{E}_{je} \in \left(-E_{maxj}, E_{maxj}\right)$, and $\overline{E}_{dqie}, \overline{E}_{bqje} \in [0, 1]$, such that $h(0) = 0$. Thus, the roots also represent the equilibrium points of the nonlinear system. Exponential stability at the origin can be investigated via system's (6.70) corresponding Jacobian matrix:

$$J_1 = \begin{bmatrix} -Hd\left[E_{qe}\right]^2[U]\,r_v^{-1} & \mathbf{0}_{(n+m)\times(n+m)} \\ \left(Hd\,[U]\,r_v^{-1} - 2k\right)[E_e]\left[E_{qe}\right]E_{max}^{-2} & -2k\left[E_{qe}\right]^2 \end{bmatrix} \tag{6.73}$$

where it is obvious that J_1 is negative definite since it is lower triangular and the diagonal elements $-Hd\left[E_{qe}\right]^2[U]\,r_v^{-1}$ and $-2kE_{qe}^2$ are diagonal and negative definite matrices.

Therefore matrix J_1 is Hurwitz. Hence, there exist $\rho_1 > 0$ and a domain $\tilde{D}_z = \left\{z \in R^{2n}, \parallel z \parallel_2 < \rho_1\right\}$ where $\tilde{D}_z \subseteq D_z$ such that (6.70) is exponentially stable at the origin uniformly in x.

To obtain the reduced model, the roots $\overline{E}$ and $\overline{E}_q$ are substituted from (6.72) into (6.66), yielding

$$\begin{bmatrix} \dot{i}_{in} \\ \dot{V} \end{bmatrix} = \begin{bmatrix} L^{-1}\left(-r_v i_{in} + \overline{E}\right) \\ FC^{-1}[V]^{-1}\left(\left([U] - \left[\overline{E}\right] + r_v\,[i_{in}]\right)i_{in} - G\right) - C^{-1}i \end{bmatrix} \tag{6.74}$$

with $F = \begin{bmatrix} 3I_n & \mathbf{0}_{n\times m} \\ \mathbf{0}_{m\times n} & I_m \end{bmatrix}$, $G = \begin{bmatrix} r_v I_q^2 & \mathbf{0}_{n\times m} \\ \mathbf{0}_{m\times n} & \mathbf{0}_{m\times m} \end{bmatrix}$, $i_{in} = \left[I_d^T\, i_L^T\right]^T$, $V = \left[V_r^T\, V_b^T\right]^T$, $i = \left[i_r^T\, i_b^T\right]^T$, $C = diag\{C_k\}$, $L = diag\{L_k\}$.

In the literature, the above model is referred to as quasi-steady-state model, since $\overline{E}$ and $\overline{E_q}$ introduce a velocity $\left[\dot{\overline{E}}_d\ \dot{\overline{E}}\right]^T = \varepsilon^{-1}g$ that is very large when ε is small and $g \neq 0$, leading to fast convergence to a root $h(I_d, I_q, i_L, V_r, V_b)$, which also represents the equilibrium of the boundary-layer.

The second equation of (6.66) is independent, thus there are n eigenvalues where $\lambda_i = -\frac{r_{vi}}{L_{si}} < 0$. The corresponding Jacobian matrix of the reduced system (6.74) that remains to be investigated will have the following form

$$J_2 = \begin{bmatrix} -L^{-1}r_v & -L^{-1}AD \\ B\left([U]+[\overline{E}]\right) - B\left[i_{ine}\right]\left([V_e]^{-1}[U]-AD\right) & -C^{-1}R^{-1}\left(I_{n+m} - \mathbf{1}_{(n+m)\times(n+m)}D\right) \end{bmatrix} \tag{6.75}$$

with matrices A and B being

$$A = H^{-1}r_v\,[U]^{-1}\,d^{-1}\mathbf{1}_{(n+m)\times(n+m)} \tag{6.76}$$

$$B = FC^{-1}\,[V_e]^{-1} \tag{6.77}$$

By virtue of Theorem 1, the characteristic polynomial can be calculated from

$$\mid \lambda I_{2(n+m)} - J_2 \mid = \mid \lambda^2 I_{n+m} + \lambda \mathbf{C}D + \mathbf{K}D \mid = 0, \tag{6.78}$$

with

$$\mathbf{C} = L^{-1}r_v D^{-1} + B\left[i_{ine}\right]\left([V_e]^{-1}[U]\,D^{-1} - A\right) + C^{-1}R^{-1}\left(D^{-1} - \mathbf{1}_{(n+m)\times(n+m)}\right) \tag{6.79}$$

$$\mathbf{K} = L^{-1}r_v\left(B\left[i_{ine}\right][V_e]^{-1}\,[U]\,D^{-1} + C^{-1}R^{-1}\left(D^{-1} - \mathbf{1}_{(n+m)\times(n+m)}\right)\right) + L^{-1}r_v H^{-1}Bd^{-1}\mathbf{1}_{(n+m)\times(n+m)} \tag{6.80}$$

Following factorisation the matrices $\mathbf{C}$ and $\mathbf{K}$ become

$$\mathbf{C} = \mathbf{Q}_1\left(\mathbf{Q}_1^{-1}\left(L^{-1}r_v D^{-1} + B\left[i_{ine}\right][V_e]^{-1}[U]D^{-1} + C^{-1}R^{-1}D^{-1}\right) - \mathbf{1}_{(n+m)\times(n+m)}\right) \tag{6.81}$$

$$\mathbf{K} = L^{-1}r_v\mathbf{Q}_2\left(\mathbf{Q}_2^{-1}\left(B\left[i_{ine}\right][V_e]^{-1}\,[U]\,D^{-1} + C^{-1}R^{-1}D^{-1}\right) - \mathbf{1}_{(n+m)\times(n+m)}\right) \tag{6.82}$$

with

$$\mathbf{Q}_1 = BH^{-1}r_v\,[U]^{-1}\,d^{-1} + C^{-1}R^{-1} \tag{6.83}$$

$$\mathbf{Q}_2 = C^{-1}R^{-1} - H^{-1}Bd^{-1} \tag{6.84}$$

Let the characteristic polynomial be

$$\mid \lambda^2 D^{-1} + \lambda\mathbf{C} + \mathbf{K} \mid\mid D \mid = 0. \tag{6.85}$$

Defining $\mathbf{Q} = L^{-1}r_v\mathbf{Q}_2$, the characteristic polynomial becomes

$$\mid \mathbf{Q} \mid\mid \lambda^2\mathbf{Q}^{-1}D^{-1} + \lambda\tilde{\mathbf{C}} + \tilde{\mathbf{K}} \mid\mid D \mid = 0. \tag{6.86}$$

with $\tilde{\mathbf{C}} = \mathbf{Q}^{-1}\mathbf{C}$ and $\tilde{\mathbf{K}} = \mathbf{Q}^{-1}\mathbf{K}$. As the determinants $\mid \mathbf{Q} \mid$ and $\mid D \mid$ are positive, the polynomial reduces to

$$\mid \lambda^2\mathbf{Q}^{-1}D^{-1} + \lambda\tilde{\mathbf{C}} + \tilde{\mathbf{K}} \mid = 0, \tag{6.87}$$

which is a quadratic eigenvalue problem (QEP) with $\tilde{\mathbf{K}}$ symmetrical, and $\tilde{\mathbf{C}}$, according to Lemma 2 in [17], diagonalisable whose eigenvalues are all real, since it is a product of a positive-definite diagonal and a symmetrical matrix.

The characteristic equation then becomes

$$|\lambda^2\mathbf{Q}^{-1}D^{-1} + \lambda\mathbf{P}\Lambda\mathbf{P}^{-1} + \tilde{\mathbf{K}}\,| = 0, \tag{6.88}$$

$$|\lambda^2\mathbf{P}^{-1}\mathbf{Q}^{-1}D^{-1}\mathbf{P} + \lambda\Lambda + \mathbf{P}^{-1}\tilde{\mathbf{K}}\mathbf{P}\,| = 0. \tag{6.89}$$

Note that Λ is a diagonal matrix with the same index of inertia as matrix $\tilde{\mathbf{C}}$, while the similarity transformations $\mathbf{P}^{-1}\mathbf{Q}^{-1}D^{-1}\mathbf{P}$ and $\mathbf{P}^{-1}\tilde{\mathbf{K}}\mathbf{P}$ are symmetrical, as $\mathbf{P}$ is unitary $(\mathbf{P}^{-1} = \mathbf{P}^T)$, and they share the same spectrum as $\mathbf{Q}^{-1}D^{-1}$ and $\tilde{\mathbf{K}}$, respectively. If $\mathbf{Q}^{-1}D^{-1}$, Λ and $\tilde{\mathbf{K}}$ are positive definite, then $Re(\lambda) < 0$ which means that J_2 is Hurwitz. Hence, since $\mathbf{Q}^{-1}D^{-1}$ is already positive-definite, it is sufficient to show that $\Lambda > 0$, or equivalently that $\tilde{\mathbf{C}}$ has positive eigenvalues, and $\tilde{\mathbf{K}} > 0$. Since matrix $\tilde{\mathbf{C}}$ is represented by a multiplication where one term is the diagonal matrix $\mathbf{Q}^{-1}\mathbf{Q}_1 > 0$, according to the same *Lemma 2* in [17], the remaining symmetrical term, denoted $\tilde{\mathbf{C}}^*$, will have the same index of inertia as $\tilde{\mathbf{C}}$. The condition $\tilde{\mathbf{C}}^* > 0$ becomes

$$\tilde{\mathbf{C}}^* = \mathbf{Q}_1^{-1}\left(L^{-1}r_vD^{-1} + B\,[i_{ine}]\,[V_e]^{-1}\,[U]\,D^{-1} + C^{-1}R^{-1}D^{-1}\right) - \mathbf{1}_{(n+m)\times(n+m)} \tag{6.90}$$

which represents a sum between a diagonal positive-definite real matrix and the real symmetric matrix $-\mathbf{1}_{(n+m)\times(n+m)}$. According to Lemma 1 in [17], if

$$\left(\frac{r_{vk}C_k}{L_k} + \frac{\alpha_kU_ki_{inke}}{V_{ke}^2} + \frac{1}{R_k}\right)\frac{1}{\lambda_{Dk}} - \left(\frac{\beta_kr_{vk}i_{inke}}{V_{ke}U_kd_k} + \frac{1}{R_k}\right)(n+m) > 0, \tag{6.91}$$

$\forall k = 1\ldots n+m$ holds, then $\tilde{\mathbf{C}}^* > 0$ is satisfied. When $k = 1\ldots n$, then $\alpha_k = 3$ and $\beta_k = 2$, whereas when $k = n+1\ldots n+m$, then $\alpha_k = \beta_k = 1$. Regarding condition $\tilde{\mathbf{K}} > 0$, taking into account Assumption 9, and according to the same Lemma 1 if

$$\left(\frac{\alpha_kU_ki_{inke}}{V_{ke}^2} + \frac{1}{R_k}\right)\frac{1}{\lambda_{Dk}} - \left(\frac{1}{R_k} - \frac{\beta_k}{V_{ke}d_k}\right)(n+m) > 0 \tag{6.92}$$

$\forall k = 1\ldots n+m$ holds, then $\tilde{\mathbf{K}} > 0$ is satisfied. Hence, if the two conditions (6.91)–(6.92) are satisfied for each converter then there exist $\rho_2 > 0$ and a domain $\tilde{D}_x = \{x \in R^{2n},\ \|x\|_2 < \rho_2\}$ where $\tilde{D}_x \subseteq D_x$ such that the reduced model is exponentially stable at the origin.

According to Theorem 4, there exists $\varepsilon^* > 0$ such that for all $\varepsilon < \varepsilon^*$ (or equivalently $c_d > \frac{1}{\varepsilon^*}I_n + \delta_d$ and $c > \frac{1}{\varepsilon^*}I_m + \delta$), the equilibrium point

$$\left[\,I_{de}^T\ I_{qe}^T\ i_{Le}^T\ V_{re}^T\ V_{be}^T\ E_{de}^T\ E_{dqe}^T\ E_{be}^T\ E_{bqe}^T\,\right]$$

of (6.69)–(6.70) with $E_{die} \in \left(-E_i^{max}, E_i^{max}\right)$, $E_{je} \in \left(-E_j^{max}, E_j^{max}\right)$ and E_{dqie}, $E_{bqje} \in (0, 1)$ is exponentially stable; thus completing the stability analysis of the entire DC microgrid.

6.2.4 Validation of Closed-Loop System Stability

In order to validate the theoretical stability analysis presented in Sect. 6.2.3 and demonstrate how conditions (6.91)–(6.92) can be tested, let us consider the system in Sect. 6.2.5 with parameters given in Table 6.2. Although (6.91)–(6.92) might seem difficult to verify, by taking into account that $E_d \in \left[-E_d^{max}, E_d^{max}\right]$, $E \in [-E^{max}, E^{max}]$ and E_{dq}, $E_{bq} \in [0, 1]$, which is guaranteed by the proposed control design, the procedure to verify whether the system is stable is the following: One can start by selecting a virtual voltage E_{de}, inside its defined range, for the rectifier. Then the values of the equilibrium points of the inductor current and load voltage are computed. Based on these obtained values, the remaining virtual voltages E_e of the DC/DC converter can be calculated. Thereafter, critical points of the output voltages are calculated, followed by the eigenvalues of matrix D. Finally, the two conditions can be tested for each converter.

Hence, following this procedure for different values of the set power of the battery, P_{setBAT}, corresponding to the battery operation, charging and discharging, respectively, one can observe in Fig. 6.8 that for any E_d in the bounded range $\left(-E_d^{max}, E_d^{max}\right) = (-21, 21)$, the expressions (6.91)–(6.92) for each converter are positive, thus ensuring closed-loop stability.

To further validate the stability analysis, in Fig. 6.9, a graphical interpretation of the stability conditions is provided for the entire range of the set power, P_{setBAT}, to visually confirm that the two stability conditions always take positive values in the entire operating range of the particular DC microgrid.

Table 6.2 Controller and system parameters of a DC microgrid consisting of a bidirectional DC/DC boost and a three-phase AC/DC converter feeding a common CPL

Parameters	Values	Parameters	Values
U_{RMS}	110 V	U_{bat}	200 V
R_{rec}	0.7 Ω	R_{bat}	1.2 Ω
L_{phase}	2.2 mH	L_{bat}	2.3 mH
C_{rec}	1200 μF	C_{bat}	2000 μF
d_{rec}	0.015	d_{bat}	0.030
P	200 W	k	1000
c_d	2.1	c_{bat}	180
r_v	7 Ω	r_{vb}	5 Ω
E_d^{max}	21 V	E_b^{max}	5 V

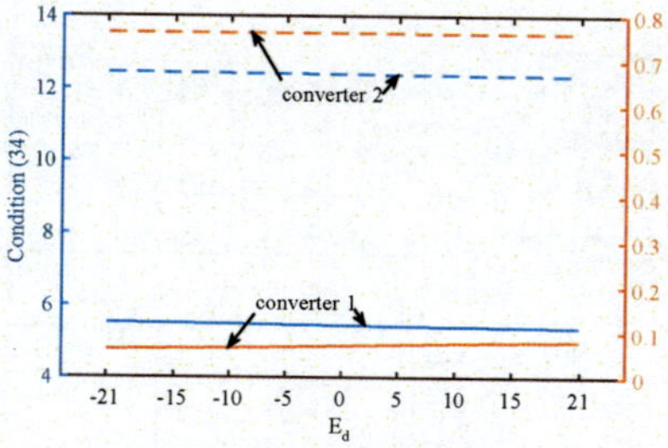

(a) $P_{setBAT} = 0\,\mathrm{W}$

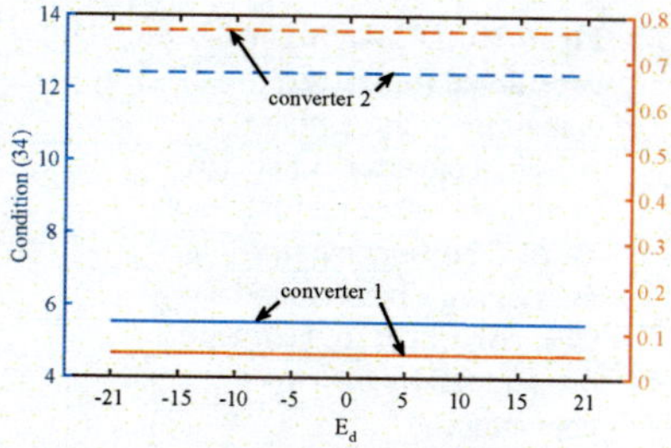

(b) $P_{setBAT} = 500\,\mathrm{W}$

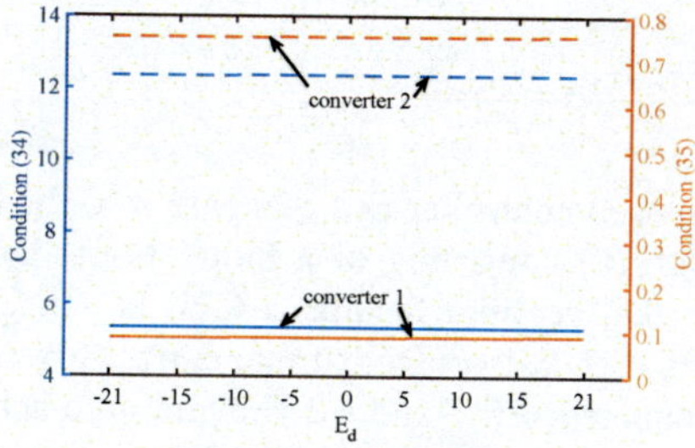

(c) $P_{setBAT} = -500\,\mathrm{W}$

Fig. 6.8 Checking stability conditions (6.91)–(6.92)

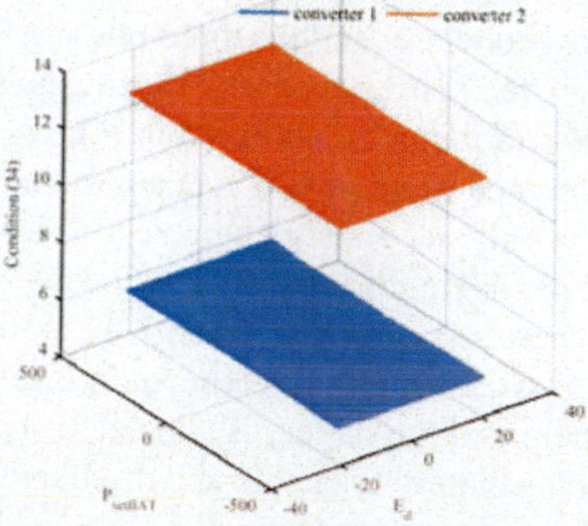

(a) 3D visualisation of condition (6.91)

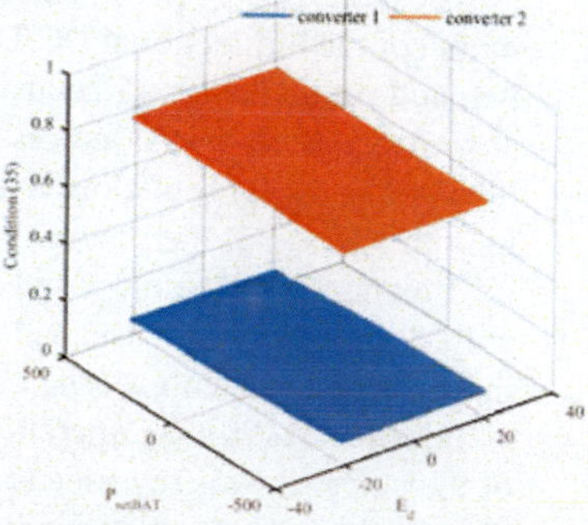

(b) 3D visualisation of condition (6.92)

Fig. 6.9 Graphical representation of the stability conditions (6.91)–(6.92)

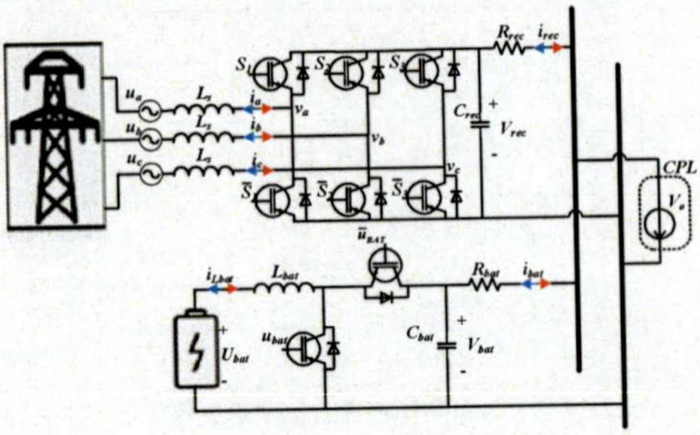

Fig. 6.10 DC microgrid considered for testing, containing a three-phase AC/DC converter connected to the grid, a bidirectional DC/DC boost converter interfacing a battery, and a CPL connected to the main bus and fed by the two converters

6.2.5 *Simulation Results*

To test the proposed controller and compare it to the cascaded PI approach, a DC microgrid (Fig. 6.10) consisting of a bidirectional DC/DC boost converter and a three-phase AC/DC rectifier feeding a CPL is considered having the parameters specified in Table 6.2, with the maximum current given by the appropriate choice of the control parameters E^{max} and r_v. The aim is to achieve tight voltage regulation around the reference value $V^* = 400$ V, accurate power sharing in a $2:1$ ratio among the paralleled AC/DC and DC/DC converters at the load bus while also assuring protection against overcurrents. But first the conditions for stability must hold. That is why Sect. 6.2.4 is vital, prior to simulating and analysing the system performance.

The model has been implemented in Matlab/Simulink, by considering the average model of the bidirectional DC/DC boost converter and the three-phase AC/DC rectifier, and simulated for 45 s considering a full testing scenario. During the first 5 s, the power requested by the load is 200 W and it can be observed in Fig. 6.11b that the load voltage V_o is kept close to the reference value of 400 V, at approximately 398 V in both cases. But the power sharing is only accurately guaranteed (Fig. 6.11c) in a $2:1$ manner with the proposed controller having $i_{BAT} \approx 0.17$ A and $i_{REC} \approx 0.34$ A, unlike the case with cascaded *PI*s where $i_{BAT} \approx 0.16$ A and $i_{REC} \approx 0.35$ A. The input currents haven't reached their imposed limits yet as shown in Fig. 6.11a.

For the next 20 s the operation principle of the battery is simulated. The direction of the power flow is reversed to allow the battery to charge and discharge. At $t = 5$ s the power set by the battery controller becomes negative $P_{setBAT} = -150$ W, thus leaving the battery to be supplied by the three-phase rectifier. The input current of the battery becomes negative, while the rectifier's input current increases to satisfy the new amount of power requested in the network (Fig. 6.11a). The power sharing ratio between the battery and the rectifier disappears since the current of the battery changes its direction, and becomes negative as shown in Fig. 6.11a. The load voltage remains closely regulated to the desired 400 V value, at around 396.5 V in both cases. After 10 s the set value of the power returns to its initial 0 value, allowing the battery to return to its former discharging state. The power sharing ratio comes back to $2:1$ as displayed in Fig. 6.11c.

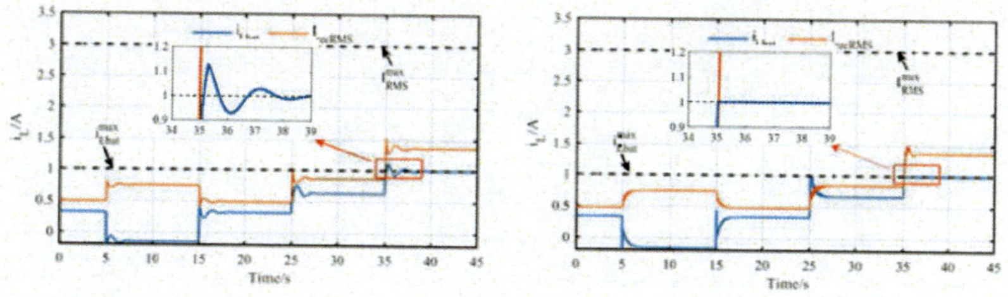

(a) inductor currents

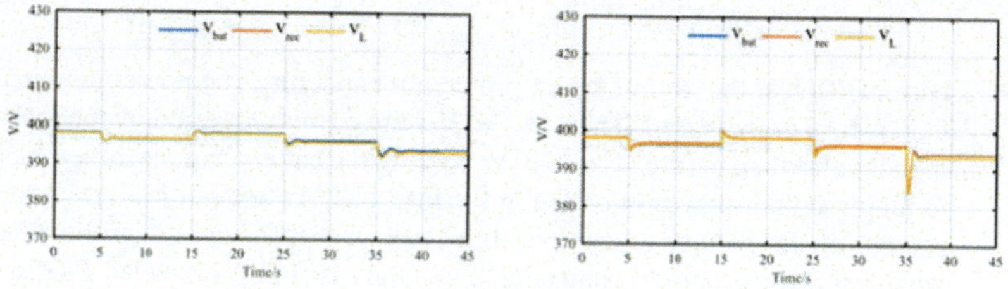

(b) output voltages

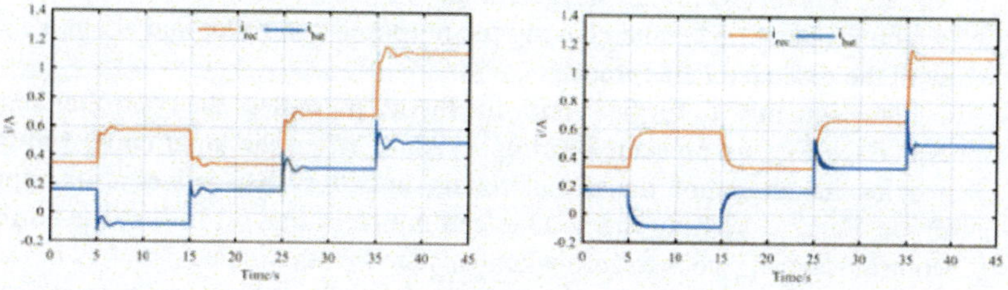

(c) output currents

Fig. 6.11 Simulation results of the DC microgrid system with *PI* cascaded control (left) and the proposed controller (right)

At $t = 25$ s the power requested by the load increases $P = 400$ W and, thus, more power is needed from the battery and the three-phase rectifier to be injected in the microgrid. The load voltage drops down to 396 V according to Fig. 6.11b when using the proposed controller and $V_{bat} = 395.5$ V when having cascaded *PI*s. At the same time, the input currents increase and, therefore, the power injected increases at the common bus (Fig. 6.11a). One can see that the sharing is kept between the two sources, the battery and rectifier, to the desired proportion of 2 : 1 having $i_{BAT} \approx 0.34$ A and $i_{REC} \approx 0.68$ A with the proposed controller, and $i_{BAT} \approx 0.32$ A and $i_{REC} \approx 0.7$ A with the cascaded *PI* technique, as presented in Fig. 6.11c, given the fact that none of the inductor currents have reached their maximum allowed

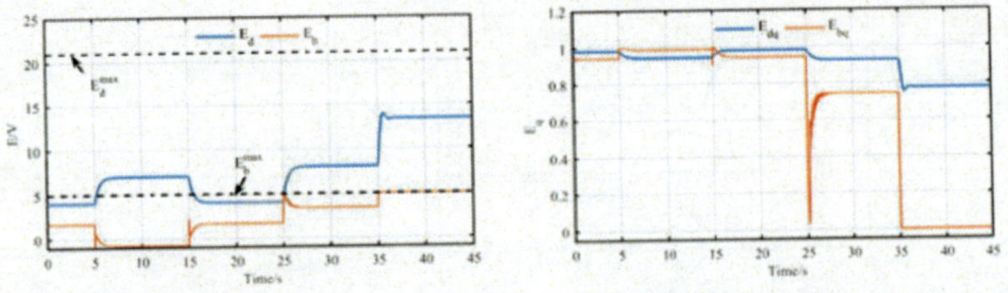

(a) Virtual voltages (b) Additional control states

Fig. 6.12 Dynamic response of the control states

current. To test the input current protection capability, the power demanded by the load is further increased. Thus, at $t = 35$ s the power requested by the load reaches a higher value than before, $P = 640$ W, forcing the battery and the three-phase rectifier to increase their power injection at the load bus. As noticed in Fig. 6.11a, the input current of the battery reaches its limit $i_{LBAT} = i_{LBAT}^{max} = 1$ A without violating it when using the proposed controller, but in the case of the cascaded PIs the transient current exceeds the upper limit prior reaching to steady-state. The power sharing is sacrificed (Fig. 6.11c) to ensure uninterruptible power supply to the load. The load voltage remains within the desired range, $V_o = 393.5$ V with a voltage drop of 6.5 V, which is about 1.5% when having the proposed controller and about $V_o = 392.5$ V with the cascaded PI approach.

Consequently, to further verify the theory presented, the controller states E, E_d and E_{dq}, E_{bq} are presented in Fig. 6.12a, b. When the input current of the battery reaches its maximum, the virtual voltage of battery also arrives at its imposed limit $E_b = E_{maxb} = i_L^{max} r_{vb} = 5$ V. One can notice in Fig. 6.12b that the corresponding control state E_{bq} goes to zero when E_b reaches maximum.

It is noted that for the particular DC microgrid scenario and the parameters used, the closed-loop performance with the cascaded PI control remains stable. However, this might not be true for a different system since there is no rigorous proof of stability. On the other hand, the proposed control approach provides a strong theoretic framework, as proven in Sect. 6.2.3, that can be easily tested for different systems as well.

6.2.6 Experimental Results

A DC microgrid, with the parameters given in Table 6.3, consisting of two parallel Texas Instruments DC/DC boost converters (see Appendix A) connected to a common DC bus and feeding an ETPS ELP-3362F electronic load, operated in CPL mode, is experimentally tested. A switching frequency of 60 kHz was used for the

Table 6.3 Controller and experimental testbed parameters of a DC microgrid consisting for two Texas Instruments DC/DC boost converters feeding an ETPS ELP-3362 electronic load acting as a CPL

Parameters	Values	Parameters	Values
U_1	36 V	U_2	24 V
R_1	2.4 Ω	R_2	3 Ω
$L_{1,2}$	0.3 mH	$C_{1,2}$	300 μF
d_1	0.2	d_2	0.4
$r_{v1,2}$	20 V	k	1000
c_1	873	c_2	655
E_1^{max}	30 V	E_2^{max}	50 V

pulse-width-modulation of both converters. The aim is to experimentally validate the proposed nonlinear current-limiting control scheme. The main tasks are to regulate the output voltage to $V^* = 48$ V and regulate the power in a 2 : 1 ratio, whilst ensuring overcurrent protection.

As one can see in Fig. 6.13a, when the power changes from 40 to 60 W, the voltage is kept close to the reference value of 48 V, while the output currents are accurately shared proportionally to the sources rating, in a 2 : 1 manner, having $i_2 \approx 0.45$ A and $i_1 \approx 0.9$ A, provided the input currents, i_{L1} and i_{L2}, have not reached their upper limit.

In Fig. 6.13b, the load power demand decreases from 60 to 40 W. The output current are accurately shared, having $i_1 \approx 0.6$ A and $i_2 \approx 0.3$ A, and the load voltage is kept fixed at 48 V.

To test the current-limiting capability, the power increases from 40 to 80 W, as displayed in Fig. 6.13c. One converter reaches to its imposed limit ($i_{L1} \approx 1.5$ A), the power sharing is sacrificed to ensure the uninterrupted power supply of the load. The load voltage is still fairly close to the rated value of 48 V. As it can be seen, the current limitation is not exactly at the 1.5 A limit. This is due to the fact that the parasitic resistance, r_{in}, of the converter's inductance is ignored, in the experiment and the analysis, which in turn causes a slightly lower bound of the input current. If the parasitic resistance is considered, then based on the ISS analysis in Sect. 6.2.2, one can easily obtain that the controller parameters E^{max} and r_v should satisfy $i_L^{max} = \frac{E^{max}}{r_v + r_{in}}$ in order to reach the upper limit of the converter. Nevertheless, it is clear that by ignoring this resistance, the current still remains below i_L^{max} as desired.

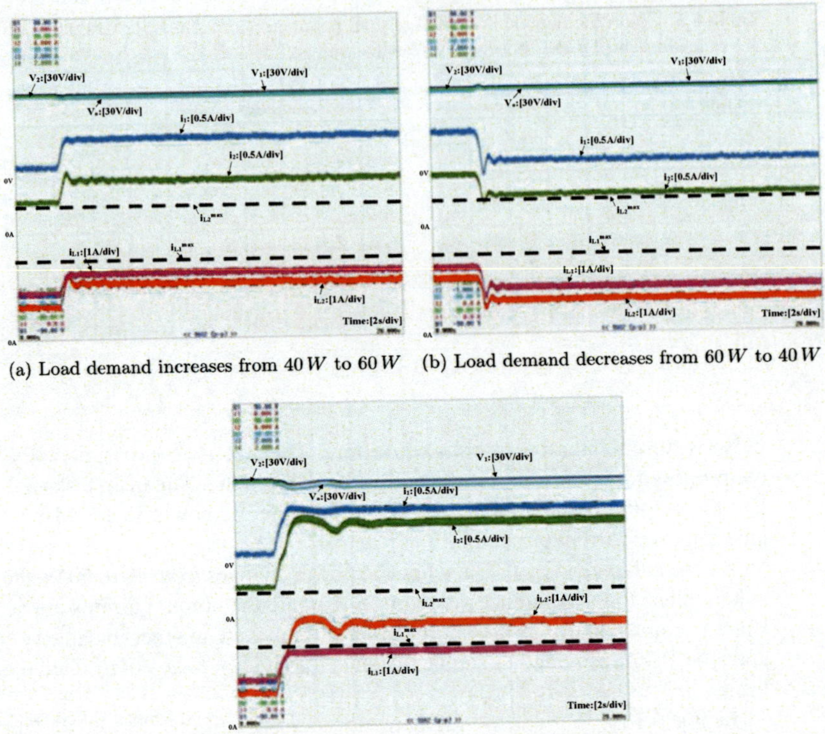

(a) Load demand increases from 40 W to 60 W (b) Load demand decreases from 60 W to 40 W

(c) Load demand increases from 40 W to 80 W

Fig. 6.13 Experimental results under the proposed controller

6.3 Conclusions

In Sect. 6.1, a detailed control design was presented for a DC microgrid framework. The nonlinear dynamic control scheme was developed to ensure reactive power control, load power sharing and output voltage regulation, with an inherent input current limitation. Introducing a virtual dynamic resistance for the three-phase rectifier and a constant virtual resistance with a bounded dynamic virtual voltage for the bidirectional DC/DC boost converter, it has been proven that the input currents of the converters will never violate a maximum given value. This feature is guaranteed without any knowledge of the system parameters and without any extra measures such as limiters or saturators, thus, addressing the issue of integrator wind-up and instability problems that often happen with the traditional overcurrent controllers' design. The effectiveness of the proposed scheme and its overcurrent capability was verified by simulating a DC microgrid considering a full testing scenario.

In Sect. 6.2, a detailed control design was presented for multiple parallel operated three-phase AC/DC and bidirectional DC/DC boost converters in a DC microgrid framework, loaded by a CPL. The nonlinear dynamic control scheme was developed to ensure load power sharing and output voltage regulation, with an inherent input current limitation. The stability of the entire DC microgrid was analytically proven when the system supplies a CPL using singular perturbation theory. Introducing a constant virtual resistance with a bounded dynamic virtual voltage for the three-phase AC/DC and for the bidirectional DC/DC boost converter, it has been shown that the input currents of each converter will never violate a maximum given value. This feature is guaranteed without any knowledge of the system parameters and without any extra measures such as limiters or saturators, thus, addressing the issue of integrator wind-up and instability problems that can occur with the traditional overcurrent controllers' design. The effectiveness of the proposed scheme and its overcurrent capability are verified by simulating a DC microgrid considering different load power variations and battery operations (charging, discharging), and by experimentally testing a parallel converter microgrid configuration feeding an electronic load, acting as a CPL.

References

1. A. Braitor, A.R. Mills, V. Kadirkamanathan, G.C. Konstantopoulos, P.J. Norman, C.E. Jones, Control of DC power distribution system of a hybrid electric aircraft with inherent overcurrent protection, in *2018 IEEE International Conference on Electrical Systems for Aircraft, Railway, Ship Propulsion and Road Vehicles International Transportation Electrification Conference (ESARS-ITEC)* (IEEE, New York, 2018), pp. 1–6
2. G.H. Gaynor, *Boeing and the 787 Dreamliner* (Wiley-IEEE Press, New York, 2015), pp. 187–218
3. E. Alcorta-Garcia, A. Zolghadri, P. Goupil, A nonlinear observer-based strategy for aircraft oscillatory failure detection: A380 case study. IEEE Trans. Aerosp. Electron. Syst. **47**(4), 2792–2806 (2011)
4. A. Ghodbane, M. Saad, C. Hobeika, J. Boland, C. Thibeault, Design of a tolerant flight control system in response to multiple actuator control signal faults induced by cosmic rays. IEEE Trans. Aerosp. Electron. Syst. **52**(2), 681–697 (2016)
5. Q.C. Zhong, G.C. Konstantopoulos, Current-limiting three-phase rectifiers. IEEE Trans. Industr. Electron. **65**(2), 957–967 (2018)
6. T.-S. Lee, Lagrangian modeling and passivity-based control of three-phase AC/DC voltage-source converters. IEEE Trans. Industr. Electron. **51**(4), 892–902 (2004)
7. R. Ortega, A. Loria, P.J. Nicklasson, H. Sira-Ramirez, *Passivity-Based Control of Euler-Lagrange Systems: Mechanical, Electrical and Electromechanical Applications* (Springer, New York, 1998)
8. A. Braitor, P.R. Baldivieso-Monasterios, G.C. Konstantopoulos, V. Kadirkamanathan, Current-limiting droop control design of paralleled AC/DC and DC/DC converters in DC micro-grids, in *IECON 2018—44th Annual Conference of the IEEE Industrial Electronics Society* (IEEE, New York, 2018), pp. 132–137
9. P. Karlsson, J. Svensson, DC bus voltage control for a distributed power system. IEEE Trans. Power Electron. **18**(6), 1405–1412 (2003)
10. J. Schonbergerschonberger, R. Duke, S.D. Round, Dc-bus signaling: a distributed control strategy for a hybrid renewable nanogrid. IEEE Trans. Industr. Electron. **53**(5), 1453–1460 (2006)
11. Q.C. Zhong, Robust droop controller for accurate proportional load sharing among inverters operated in parallel. IEEE Trans. Industr. Electron. **60**(4), 1281–1290 (2013)

12. Z. Shuai, D. He, J. Fang, Z.J. Shen, C. Tu, J. Wang, Robust droop control of DC distribution networks. IET Renew. Power Gener. **10**(6), 807–814 (2016)
13. A. Braitor, G.C. Konstantopoulos, V. Kadirkamanathan, Stability analysis and nonlinear current-limiting control design for DC micro-grids with CPLs. IET Smart Grid **3**(3), 355–366 (2020)
14. A.-C. Braitor, G.C. Konstantopoulos, V. Kadirkamanathan, Admittance matrix computation and stability analysis of droop controlled DC micro-grids with CPL. IFAC-PapersOnLine **53**(2), 13525–13530 (2020)
15. G.C. Konstantopoulos, Q.C. Zhong, B. Ren, M. Krstic, Bounded integral control of input-to-state practically stable nonlinear systems to guarantee closed-loop stability. IEEE Trans. Autom. Control **61**(12), 4196–4202 (2016)
16. H.K. Khalil, *Nonlinear Systems*, 3rd edn. (Pearson, 2014)
17. M. Su, Z. Liu, Y. Sun, H. Han, X. Hou, Stability analysis and stabilization methods of DC microgrid with multiple parallel-connected DC-DC converters loaded by CPLs. IEEE Trans. Smart Grid **9**(1), 132–142 (2018)

Chapter 7
Stability Analysis of DC Microgrids Under Decentralised Primary and Distributed Secondary Control

This chapter investigates the stability of a DC microgrid with a CPL, under decentralised primary and distributed secondary control scheme to achieve accurate power sharing and voltage restoration. At the primary layer, following the same control concept developed in the previous chapter, and to simplify the current-limiting control dynamics, one can consider the following expressions for E and E_q

$$E = E_{max} \sin \sigma \tag{7.1}$$

$$E_q = \cos \sigma. \tag{7.2}$$

By taking the time derivative, one obtains

$$\dot{E} = E_{max} \cos \sigma \dot{\sigma} \tag{7.3}$$

$$\dot{E}_q = -\sin \sigma \dot{\sigma}. \tag{7.4}$$

Replacing E and E_q in Eqs. (6.63)–(6.64), with their expressions from (7.3)–(7.4), it yields

$$E_{max} \cos \sigma \dot{\sigma} = cF(\cdot) \cos^2 \sigma \tag{7.5}$$

$$-\sin \sigma \dot{\sigma} = -cF(\cdot) \frac{E_{max} \sin \sigma \cos \sigma}{E_{max}^2}. \tag{7.6}$$

where $F(\cdot)$ represents the function that needs to be regulated to zero (e.g. droop function). Multiplying Eq. (7.5) with $\cos^{-1} \sigma$, and Eq. (7.6) with $\sin^{-1} \sigma$ one notices that the remaining equations are equivalent, leading to the expression

$$\dot{\sigma} = cF(\cdot) \frac{\cos \sigma}{E_{max}}. \tag{7.7}$$

A.-C. Braitor, *Advanced Hierarchical Control and Stability Analysis of DC Microgrids*, Springer Theses, https://doi.org/10.1007/978-3-030-95415-4_7

As a result, the two-state controller can be reduced to a single state, making it easier to work with in the theoretical analysis of hierarchical control frameworks.

Hence, at the primary control layer, the droop control concept is suitably formulated and implemented using the recently proposed state-limiting PI controller [1], which uses the simplified dynamics (7.7) to accomplish an inherent current limitation for each converter and simultaneously facilitate the stability analysis. Using limited information of the load voltage and the injected power only from neighbouring converters, a distributed secondary controller is formulated to enhance the power sharing and accurately regulate the voltage to the rated value. By analysing for the first time both the dynamics of the converters with the CPL and the two-layer control, singular perturbation theory is applied to analytically prove the stability of the entire DC microgrid. Simulation and experimental testings are performed to confirm the effectiveness and validity of the proposed method.

7.1 Dynamic Modelling of the DC Microgrid

A typical islanded DC microgrid is depicted in Fig. 7.1a, consisting of n bidirectional DC/DC boost converters connected in parallel to a DC bus and feeding a common load (CPL). Every converter includes a boosting inductor L_i, a smoothing capacitor C_i, while U_i is the DC input voltage and R_i, the output line resistance, where $i \in \mathcal{I}$. One can see, in Fig. 7.1b, the mapping of a cyber network to a physical DC microgrid. The vertices represent converters, and the edges represent the communication links for information exchange. In achieving global synchronisation, the communication graph must have at least one spanning tree. Hence, the communication network is represented by an undirected, connected and weighted graph, and the inertia of its Laplacian matrix $\mathcal{L}$, is $i(\mathcal{L}) = [n-1 \;\; 0 \;\; 1]$. Diffusive coupling (also referred to as nearest-neighbour coupling) is the most common type of coupling in distributed communication networks.

The system's nonlinear dynamic model can be described by employing Kirchhoff laws and average analysis [2], leading to the following differential equations:

$$L_i \dot{i}_{Li} = U_i - (1 - u_i)V_i \tag{7.8}$$

$$C_i \dot{V}_i = (1 - u_i)i_{Li} - i_i \tag{7.9}$$

where u_i is the duty-ratio (control) input, bounded in the range [0, 1], i_{Li} is the inductor current and V_i, i_i are the converter output voltage and current, respectively.

Rewriting the system (7.8)–(7.9) in a matrix form, the DC microgrid system takes the following form

$$\dot{i}_L = L^{-1}\left(U - (I_n - u)V\right) \tag{7.10}$$

$$\dot{V} = C^{-1}\left((I_n - u)i_L - i\right) \tag{7.11}$$

Fig. 7.1 The structure of a DC microgrid system with communication

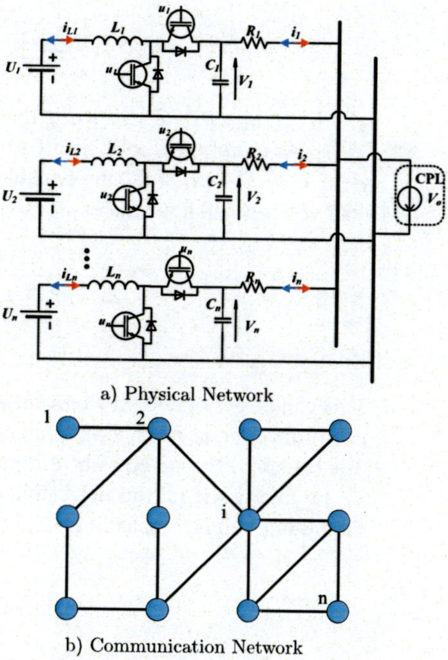

where $U = [U_1...U_n]^T$, $u = diag\{u_i\}$, $V = [V_1...V_n]^T$, $i_L = [i_{L1}...i_{Ln}]^T$, $i = [i_1...i_n]^T$, $L = diag\{L_i\}$ and $C = diag\{C_i\}$. It is clear that system (7.10)–(7.11) is nonlinear, since the control input u is multiplied with the system states $\left[i_L^T \; V^T \right]^T$.

The power balance equation for the CPL can be written as

$$P = V_o \sum_{i=1}^{n} i_i \tag{7.12}$$

where P is constant and represents the load power demand and V_o represents the load voltage. One can notice, that the output currents have the following expression $i_i = \frac{V_i - V_o}{R_i}$. Then, let Assumption 1 hold.

Hence, the load voltage, V_o, is given by the real solutions of a second-order polynomial similar to Eq. (4.7) as

$$V_o = \frac{\sum_{i=1}^{n} \frac{V_i}{R_i} \pm \sqrt{\left(\sum_{i=1}^{n} \frac{V_i}{R_i}\right)^2 - 4P \sum_{i=1}^{n} \frac{1}{R_i}}}{2 \sum_{i=1}^{n} \frac{1}{R_i}}.$$

Now, let also Assumption 2 hold. Note that based on Assumptions 1 and 2, the real solution (4.7) exists if

$$\left(\sum_{i=1}^{n}\frac{U_i}{R_i}\right)^2 > 4P\sum_{i=1}^{n}\frac{1}{R_i},$$

which can be easily tested using the constant system parameters.

Let Assumption 3 hold. Similar to [3], according to Assumptions 2 and 3, which yield $V_i \geq U_i - i_{Li}^{max} R_i$, the feasible solution is the high voltage in (4.7), thus, the load voltage will have the expression shown previously in (4.8), i.e.

$$V_o = \frac{\sum_{i=1}^{n}\frac{V_i}{R_i} + \sqrt{\left(\sum_{i=1}^{n}\frac{V_i}{R_i}\right)^2 - 4P\sum_{i=1}^{n}\frac{1}{R_i}}}{2\sum_{i=1}^{n}\frac{1}{R_i}}.$$

This can be guaranteed if a current-limiting controller with $|i_{Li}| \leq i_{Li}^{max}$ is applied at each converter, as it will be explained in the sequel, and the inequality in Assumption 3 is satisfied. For an equilibrium point (i_{Lie}, V_{ie}) given by a constant control input u_i, by taking the partial derivative of the output current $i_i = \frac{V_i - V_o}{R_i}$ with respect to the output voltage V_i, as in [3–5] we obtain the admittance matrix similar to (4.11), as

$$Y = R^{-1}\left(I_n - \mathbf{1}_{n\times n} D\right),$$

with $R = diag\{R_i\}$ and D from (4.12) having the following expression

$$D = diag\left\{\frac{\partial V_o}{\partial V_i}\right\} = \begin{bmatrix} \frac{\partial V_o}{\partial V_1} & \dots & 0 \\ \vdots & \ddots & \vdots \\ 0 & \dots & \frac{\partial V_o}{\partial V_n} \end{bmatrix} =$$

$$= \frac{1}{2\sum_{i=1}^{n}\frac{1}{R_i}}\left(R^{-1} + \frac{\sum_{i=1}^{n}\frac{V_i}{R_i}}{\sqrt{\left(\sum_{i=1}^{n}\frac{V_i}{R_i}\right)^2 - 4P\sum_{i=1}^{n}\frac{1}{R_i}}} R^{-1}\right),$$

where $\sqrt{\left(\sum_{i=1}^{n}\frac{V_i}{R_i}\right)^2 - 4P\sum_{i=1}^{n}\frac{1}{R_i}} > 0$ according to Assumption 1. As R is a diagonal positive-definite matrix, one can notice that matrix D is also a diagonal positive-definite matrix, with the eigenvalues of the form

$$\lambda_i(D) = \frac{1}{2\sum_{i=1}^{n}\frac{1}{R_i}}\left(\frac{1}{R_i} + \frac{\sum_{i=1}^{n}\frac{V_i}{R_i}}{\sqrt{\left(\sum_{i=1}^{n}\frac{V_i}{R_i}\right)^2 - 4P\sum_{i=1}^{n}\frac{1}{R_i}}}\frac{1}{R_i}\right),$$

$\forall i \in \mathcal{I}$.

7.2 Proposed Controller Design

7.2.1 *Primary Control Steady-State Analysis*

When multiple converters are connected in parallel in a DC microgrid, power sharing without the need of communication among the different converters, in a decentralised manner, is often achieved via droop control in different forms: robust droop [6], nonlinear droop [7], quadratic droop [8], inverse-droop [9], dead-band droop [10], adaptive droop [11], or other variations of droop control strategies [12, 13]. In dynamic form, the conventional droop control becomes

$$\dot{V}_i = V^* - V_i - m_i P_i. \tag{7.13}$$

where P_i is the injected power of the i-th converter, m_i is the positive droop coefficient and V^* is the rated/nominal voltage. By further looking into (7.13), it is clear that the output voltage V_i will deviate from the nominal voltage V^* as long as $P_i \neq 0$. Furthermore, the larger the droop gain m_i, the more the voltage deviation $V^* - V_i$ becomes.

In order to regulate the output voltage V_i to the nominal value V^*, and at the same time maintain the power sharing accuracy, a correction term, e_i is added into the droop function (7.13), as

$$\dot{V}_i = V^* - V_i - m_i P_i + e_i, \tag{7.14}$$

where e_i is obtained from the dynamics of the secondary control layer. The first task in this section is to design a primary controller that inherits the droop control concept and additionally maintains an upper limitation for the input current for each converter independently of the system parameters. To this end, the duty-ratio input of each boost converter is proposed to take the form

$$u_i = 1 - \frac{r_{vi} i_{Li} + U_i - E_{maxi} \sin \sigma_i}{V_i}, \tag{7.15}$$

where r_{vi} represents a constant virtual resistance and E_{maxi} a constant maximum virtual voltage for the i-th converter, chosen to satisfy, in matrix form, $I_{max} = E_{max} r_v^{-1}$. Inspired by the state-limiting PI (sl-PI) controller proposed in [1], σ_i is designed to follow the nonlinear dynamics:

$$\dot{\sigma}_i = \frac{k_i}{E_{maxi}} \left(V^* - V_i - m_i \frac{U_i E_{maxi} \sin \sigma_i}{r_{vi}} + e_i \right) cos \sigma_i \tag{7.16}$$

with k_i, E_{maxi} being positive constants. Based on the sl-PI controller analysis in [1], it is proved that the proposed structure of the control dynamics guarantees a given bound for σ_i, i.e. $\sigma_i \in \left[-\frac{\pi}{2}, \frac{\pi}{2}\right]$, in addition to the current limitation $|i_{Li}| \leq i_{Li}^{max}$.

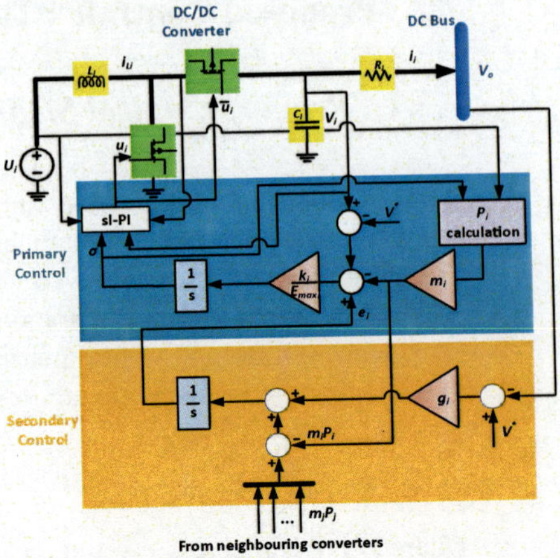

Fig. 7.2 Detailed diagram with primary and secondary controller

For more details on the boundedness of σ_i and i_{Li}, the reader is referred to [1]. A detailed diagram of the control implementation is shown in Fig. 7.2. Note that the droop function in the proposed control dynamics (7.16) differs from the conventional one in (7.14), since the term P_i has been replaced with $\frac{U_i E_{maxi} sin \sigma_i}{r_{vi}}$. In order to explain why the new term represents the converter power at the steady-state, let's replace the control input, u_i, from (7.15) into (7.8). This results in the closed-loop current dynamics

$$L_i \dot{i}_{Li} = -r_{vi} i_{Li} + E_{maxi} \sin \sigma_i, \tag{7.17}$$

where one can observe that $E_{maxi} \sin \sigma_i$ represents a virtual voltage, and r_{vi} a virtual resistance. One can see that the current i_{Li} dynamics of each converter are partially decoupled from the voltages V_i. At steady state there is

$$i_{Li} = \frac{E_{maxi} \sin \sigma_i}{r_{vi}}, \tag{7.18}$$

and since $P_i = U_i i_{Li}$, equation (7.16) has incorporated the expression

$$P_i = \frac{U_i E_{maxi} \sin \sigma_i}{r_{vi}} \tag{7.19}$$

which will represent the converter power at the steady-state. This new primary control structure has been proposed to facilitate the stability analysis of the entire microgrid, as it will be further explained in Sect. 7.3.

7.2.2 Secondary Control Design and Analysis

The second task in this section is to design a secondary controller that restores the DC bus voltage V_o to the reference value V^*, i.e.,

$$\lim_{t\to\infty}\left(V^* - V_o\right)(t) = 0 \tag{7.20}$$

while meeting the power sharing requirements, i.e.,

$$\lim_{t\to\infty}\sum_{j\in\aleph_i}\left(m_j P_j(t) - m_i P_i(t)\right) = 0 \tag{7.21}$$

where $\aleph_i \subset \mathcal{V}\left(\nu_j, \nu_i\right) \in \mathcal{E}$ denotes the neighbourhood set of the ith vertex of the cyber network.

Then the distributed secondary control that generates the correction term e_i for the primary controller can be designed in the dynamic form

$$\dot{e}_i = \alpha g_i\left(V^* - V_o\right) + \beta\sum_{j\in\aleph_i}\left(m_j P_j - m_i P_i\right) \tag{7.22}$$

where $\alpha, \beta \in \mathbb{R}^+$ are constant gains, $g_i = \{0, 1\}$, and P_i, P_j given from (7.19). Driven by the concept of the pinning control, since the DC bus voltage V_o may not be known by all the DGs, the pinning gain g_i is introduced, being nonzero for the DG that has access to the DC bus voltage V_o. By applying the secondary controller at each converter, (7.22) can be written in the matrix form

$$\dot{e} = \alpha g\left(V^* - V_o\right)\mathbf{1}_n - \beta\mathcal{L}mP \tag{7.23}$$

where $g = diag\{g_i\}$, $m = diag\{m_i\}$, and $P = [P_1 \ldots P_n]$.

At the steady state there is

$$\alpha g\left(V^* - V_o\right)\mathbf{1}_n - \beta\mathcal{L}mP = \mathbf{0}_n \tag{7.24}$$

Corollary 2 *Since $m_i \neq 0$, and assuming that at least one converter measures the load voltage, i.e. $\sum_{i=1}^{n} g_i > 0$, the following equations hold:*

$$m_1 P_1 = m_2 P_2 = \ldots = m_n P_n \tag{7.25}$$

$$V_o = V^* \tag{7.26}$$

Proof As the Laplacian $\mathcal{L}$ is balanced and symmetric, $ker\left(\mathcal{L}\right) = span\left(\mathbf{1}_n\right)$. That is $\mathbf{1}_n^T\mathcal{L} = \mathbf{0}_n^T$. Therefore, by left multiplication with $\mathbf{1}_n^T$, Eq. (7.24) becomes

$$\alpha\mathbf{1}_n^T g\left(V^* - V_o\right) - \beta\mathbf{1}_n^T\mathcal{L}mP = \mathbf{0}_n. \tag{7.27}$$

As the right term in (7.27) is $\mathbf{0}_n$, equality (7.26) obviously holds if $\sum_{i=1}^{n} g_i \neq 0$, or equivalently $\sum_{i=1}^{n} g_i > 0$, given that $g_i \geq 0$, $\forall i \in \mathcal{I}$. By substituting (7.26) into (7.24), it yields

$$\beta \mathcal{L} m P = \mathbf{0}_n. \tag{7.28}$$

One can easily see that $mP \in span\,(\mathbf{1}_n)$, since it is equivalent to (7.25). This completes the proof. □

7.3 Stability Analysis

7.3.1 *Closed-Loop System*

By applying the proposed controller (7.15)–(7.16), (7.22) into the DC microgrid dynamics (7.8)–(7.9), the closed-loop system can be written in the following matrix form

$$\begin{bmatrix} \dot{i}_L \\ \dot{V} \end{bmatrix} = \begin{bmatrix} L^{-1}\left(-r_v i_L + E_{max}\mathbf{sin}\,(\sigma)\right) \\ C^{-1}\,[V]^{-1}\left(r_v\,[i_L] + [U] - E_{max}\,[\mathbf{sin}\,(\sigma)]\right) i_L - i \end{bmatrix} \tag{7.29}$$

$$\begin{bmatrix} \dot{\sigma} \\ \dot{e} \end{bmatrix} = \begin{bmatrix} E_{max}^{-1} k\,[\mathbf{cos}\,(\sigma)]\left(V^* - V - [r_v]^{-1} m\,[U]\,E_{max}\mathbf{sin}\,(\sigma) + e\right) \\ \alpha g\,(V^* - V_o)\,\mathbf{1}_n - \beta \mathcal{L} m P \end{bmatrix} \tag{7.30}$$

where $\sigma = [\sigma_1 \ldots \sigma_n]^T$, $k = diag\{k_i\}$, $E_{max} = diag\{E_{maxi}\}$.

For the closed-loop system, consider the following assumption:

Assumption 10 For a constant $\sigma_{ie} \in \left(-\frac{\pi}{2}, \frac{\pi}{2}\right)$, satisfying (7.25), there exists a unique equilibrium point $(i_{Lie}, V_{ie}, \sigma_{ie}, e_{ie})$, corresponding to the desired voltage regulation (7.26).

Note that the previous assumption is made since the proof of the existence of a unique equilibrium point for a microgrid with primary and secondary control is a non-trivial problem, as one can see from [8, 14].

Since r_{vi} and k_i represent control parameters, they can be suitably selected in order for the terms $\frac{L_i}{r_{vi}}$ and $\frac{1}{k_i}$ to be sufficiently small. Hence, by introducing the perturbation parameter ε, as $\varepsilon = min\{\frac{1}{k_i}, \frac{L_i}{r_{vi}}\}$, one can conclude that there exist $\delta_L = diag\,\{\delta_{Li}\} \geq 0$ and $\delta_\sigma = diag\,\{\delta_{\sigma i}\} \geq 0$ such that $L r_v^{-1} = \frac{1}{\varepsilon} I_n + \delta_L$ and $k = \frac{1}{\varepsilon} I_n + \delta_\sigma$. Hence, (7.30) becomes

$$\begin{bmatrix} \varepsilon \dot{i}_L \\ \varepsilon \dot{\sigma} \end{bmatrix} = \begin{bmatrix} (I_n + \varepsilon \delta_L)\left(-i_L + I_{max}\mathbf{sin}\,(\sigma)\right) \\ (I_n + \varepsilon \delta_\sigma)\,E_{max}^{-1}\,[\mathbf{cos}\,(\sigma)]\left(V^* - V - m\,[U]\,I_{max}\mathbf{sin}\,(\sigma) + e\right) \end{bmatrix} \tag{7.31}$$

where, it has been taken into account that $I_{max} = r_v^{-1} E_{max}$. Therefore, the closed-loop system Eqs. (7.29)–(7.30) can be written in the following form

$$\dot{x} = f(x, z) \tag{7.32}$$
$$\varepsilon\dot{z} = q(x, z, \varepsilon) \tag{7.33}$$

where $x = \begin{bmatrix} V - V_e \\ e - e_e \end{bmatrix}$ and $z = \begin{bmatrix} i_L - i_{Le} \\ \sigma - \sigma_e \end{bmatrix}$. System (7.32)–(7.33) can then be investigated as a singularly perturbed system using two-time-scale analysis [15]. Since it represents the immediate vicinity of a bounding surface, system (7.31) is also referred to as the boundary layer, and it is analysed in the section below.

7.3.2 Boundary Layer Stability Analysis

Let functions f, q be continuously differentiable in the domain $(x, z, \varepsilon) \in D_x \times D_z \times [0, \varepsilon_0]$. Considering the scenario where the controller parameters r_{vi} and k_i are selected sufficiently large, then $\varepsilon \to 0$ and, according to the singular perturbation theory, function q will have an algebraic form of $0 = q(x, z)$ as follows

$$\begin{bmatrix} \mathbf{0}_n \\ \mathbf{0}_n \end{bmatrix} = \begin{bmatrix} -i_L + I_{max}\mathbf{sin}\,(\sigma) \\ E_{max}^{-1}\,[\mathbf{cos}\,(\sigma)]\,(V^* - V - m\,[U]\,I_{max}\mathbf{sin}\,(\sigma) + e) \end{bmatrix} \tag{7.34}$$

The roots of the above system can be computed as

$$\begin{bmatrix} i_L \\ \sigma \end{bmatrix} = \begin{bmatrix} I_{max}\mathbf{sin}\,(\sigma) \\ \mathbf{sin}^{-1}\left([U]^{-1}\,I_{max}^{-1}\,(V^* I_n - [V] + [e])\right) \end{bmatrix} \tag{7.35}$$

and can also be referred to as $z = h(x)$ with $\sigma_i \in (-\frac{\pi}{2}, \frac{\pi}{2})$, such that $h(0) = 0$. Thus, the roots also represent the equilibrium points of the nonlinear system (7.29)–(7.30). Exponential stability at the origin can be investigated via its corresponding Jacobian matrix:

$$J_1 = \begin{bmatrix} -I_n & I_{max}\,[\mathbf{cos}\,(\sigma)] \\ \mathbf{0}_{n\times n} & -[\mathbf{cos}\,(\sigma)]^2\,E_{max}^{-1} m\,[U]\,I_{max} \end{bmatrix} \tag{7.36}$$

As one can observe matrix J_1 is Hurwitz, since J_1 is upper triangular and all its diagonal elements are negative. Hence, there exist $\rho_1 > 0$ and a domain $\tilde{D}_z = \{z \in R^{2n}, \| z \|_2 < \rho_1\}$ where $\tilde{D}_z \subseteq D_z$ such that (7.33) is exponentially stable at the origin uniformly in x.

7.3.3 Reduced Model

To obtain the reduced model, the roots i_L and σ are substituted from (7.35) into (7.29)–(7.30), yielding

$$\dot{V} = C^{-1}[V]^{-1} m^{-1} \left(V^* \mathbf{1}_n - V - e\right) - C^{-1} i \tag{7.37}$$

$$\dot{e} = \alpha g \left(V^* - V_o\right) \mathbf{1}_n - \beta L \left(V^* \mathbf{1}_n - V + e\right) \tag{7.38}$$

often referred to as the quasi-steady-state model, because i_L and σ introduce a velocity $\left[\dot{i}_L\, \dot{\sigma}\right]^T = \varepsilon^{-1} q$ which is very large when ε is small and $q \neq 0$. This leads to rapid convergence to a root $h(V, e)$, which is also the equilibrium of the boundary-layer. The corresponding Jacobian matrix of the reduced model will have the form of J_2, as shown below:

$$J_2 = \begin{bmatrix} -C^{-1}[V_e]^{-2} m^{-1} (V^* I_n + [e]) - C^{-1} R^{-1} (I_n - \mathbf{1}_{n\times n} D) & C^{-1}[V_e]^{-1} m^{-1} \\ -\alpha g \mathbf{1}_{n\times n} D + \beta \mathcal{L} & -\beta \mathcal{L} \end{bmatrix}. \tag{7.39}$$

The Jacobian J_2 can be rewritten as a sum of two matrices, i.e., $J_2 = \alpha J_3 - \beta J_4$, with J_3 being

$$J_3 = \underbrace{\begin{bmatrix} \frac{1}{\alpha} C^{-1} R^{-1} & -g \\ -g & \frac{1}{\alpha} I_n \end{bmatrix}}_{X_1} \underbrace{\begin{bmatrix} \mathbf{1}_{n\times n} & \mathbf{0}_{n\times n} \\ \mathbf{0}_{n\times n} & \mathbf{0}_{n\times n} \end{bmatrix}}_{X_2} \underbrace{\begin{bmatrix} D & \mathbf{0}_{n\times n} \\ \mathbf{0}_{n\times n} & D \end{bmatrix}}_{Q} \tag{7.40}$$

and J_4 having the following expression

$$J_4 = \underbrace{\begin{bmatrix} \frac{1}{\beta} C^{-1}[V_e]^{-1} m^{-1} & \mathbf{0}_{n\times n} \\ \mathbf{0}_{n\times n} & \mathcal{L} \end{bmatrix}}_{X_3} \underbrace{\begin{bmatrix} [V_e]^{-1}(V^* I_n + [e]) + [V_e] m R^{-1} & -I_n \\ -I_n & I_n \end{bmatrix}}_{X_4}. \tag{7.41}$$

Hence, the stability problem becomes a standard eigenvalue problem (SEP) as follows

$$J_2 v = \lambda v \tag{7.42}$$

which gives

$$(\alpha X_1 X_2 Q_1 - \beta X_3 X_4)\, v = \lambda v \tag{7.43}$$

where $\lambda \in \mathbb{R}$ is an eigenvalue and $v \in \mathbb{R}^n$ is the associated eigenvector. Let $y \triangleq Q_1 v$; then the SEP becomes $\overline{J}_2 y = \lambda y$, as follows

$$\left(\alpha \underbrace{X_1 X_2}_{\overline{J}_3} - \beta \underbrace{X_3 X_4 Q_1^{-1}}_{\overline{J}_4} \right) y = \lambda y \tag{7.44}$$

Proposition 1 *Matrix* $\overline{J}_3 = X_1 X_2$ *is semi-positive stable and diagonalisable if*

$$\alpha < C^{-\frac{1}{4}} R^{-\frac{1}{4}}. \tag{7.45}$$

for all $i \in \mathcal{I}$, *where* $g_i = 1$.

Proof Matrix X_2 is symmetric and singular, with $2n-1$ eigenvalues equal to zero, i.e. $\lambda_{1...2n-1}(X_2) = 0$, and one positive eigenvalue equal to the trace of the matrix, i.e. $\lambda_{2n}(X_2) = Tr(X_2) = n$. Hence X_2 is positive semi-definite.

To prove that the symmetric matrix X_1 is positive-definite, one can use the quadratic eigenvalue problem (QEP) for matrix $-X_1$. This yields

$$|\lambda I_{2n} + X_1| = |\lambda^2 \mathbb{M} + \lambda \mathbb{C} + K| = 0 \tag{7.46}$$

where $\mathbb{M} = I_n \succ 0$ and

$$\mathbb{C} = \frac{1}{\alpha}\left(C^{-1}R^{-1} + I_n\right) \tag{7.47}$$

$$\mathbb{K} = \frac{1}{\alpha^2} C^{-1} R^{-1} - \alpha^2 g^2 \tag{7.48}$$

If $\mathbb{M}$, $\mathbb{C}$ and $\mathbb{K}$ are positive-definite, then $Re(\lambda) < 0$, and $-X_1$ is negative-definite, thus X_1 is positive-definite. The first condition $\mathbb{C} \succ 0$ is easily satisfied, given that $\alpha > 0$. Regarding the second condition $\mathbb{K} \succ 0$, since it consists of a sum of two diagonal matrices, using Lemma 1 in [16], the condition, in scalar form, becomes

$$\frac{1}{\alpha^2 C_i R_i} - \alpha^2 g_i^2 > 0, \forall i \in \mathcal{I} \tag{7.49}$$

with the pinning control gain either 0 or 1. Assuming a worst case scenario, inequality (7.45) must hold.

As X_1 is a positive definite symmetric matrix, and X_2 is a positive semi-definite symmetric matrix, then according to Lemma 3,

(1) $X_1 X_2$ (or $X_2 X_1$) is diagonalisable, i.e. $P_1^{-1} \overline{J}_3 P_1 = \Lambda_1$, with P_1 unitary and Λ_1 diagonal having the same index of inertia as matrix $\overline{J}_3$, and
(2) the eigenvalues of $X_1 X_2$ are real, and $X_1 X_2$ has the same number of positive (zero, or negative) eigenvalues with matrix X_2.

Hence, the proof of Proposition 1 is complete. □

Proposition 2 *Matrix* $\overline{J}_4 = X_3 X_4 Q_1^{-1}$ *is diagonalisable and semi-positive stable.*

Proof Matrix $\overline{J}_4$ can be split into a product of two symmetric matrices, X_3 and $X_4 Q_1^{-1}$. The latter symmetric matrix, $X_4 Q_1^{-1}$, has the following shape

$$X_4 Q_1^{-1} = \begin{bmatrix} [V_e]^{-1}(V^* I_n + [e]) D^{-1} + [V_e] m R^{-1} D^{-1} & -D^{-1} \\ -D^{-1} & D^{-1} \end{bmatrix}$$

By applying the QEP theory again for the matrix $-X_4 Q_1^{-1}$, we get

$$|\lambda I_n + X_4 Q_1^{-1}| = |\lambda^2 \mathbb{N} + \lambda \mathbb{E} + \mathbb{L}| = 0 \tag{7.50}$$

where $\mathbb{N} = I_n \succ 0$ and

$$\mathbb{E} = [V_e]^{-1}(V^* I_n + [e])D^{-1} + [V_e] m R^{-1} D^{-1} + D^{-1} \tag{7.51}$$
$$\mathbb{L} = [V_e]^{-1}(V^* I_n + [e])D^{-2} + [V_e] m R^{-1} D^{-2} - D^{-2} \tag{7.52}$$

If the conditions $\mathbb{N} \succ 0$, $\mathbb{E} \succ 0$ and $\mathbb{L} \succ 0$ hold, then $Re\,(\lambda) < 0$, and $-X_4 Q_1^{-1}$ is negative-definite, thus $X_4 Q_1^{-1}$ is positive-definite. Condition $\mathbb{E} \succ 0$ is satisfied as it represents a positive-definite diagonal matrix. Condition $\mathbb{L} \succ 0$ will hold, according to Lemma 1, if the following condition in scalar form is guaranteed

$$\left(\frac{1}{V_{ie}}\left(V^* + e_{ie}\right) + \frac{V_{ie} m_i}{R_i} - 1\right)\frac{1}{\lambda_{Di}^2} > 0, \forall i \in \mathcal{I} \tag{7.53}$$

which is easily satisfied since $\frac{(V^*+e_{ie})}{V_{ie}} > 1$ from the droop equation (7.14) at the steady-state (every converter is feeding the load, i.e. $P_i > 0$). Thus, $X_4 Q_1^{-1} \succ 0$, and since $X_3 \succeq 0$, then $\overline{J}_4$ is diagonalisable, according to Lemma 3, i.e. $P_2^{-1} \overline{J}_4 P_2 = \Lambda_2$, with P_2 unitary and Λ_2 diagonal having the same index of inertia as matrix $\overline{J}_4$.

This completes the proof of Proposition 2. □

Now let the following similarity transformation $\tilde{J}_2 = (X_4 Q_1^{-1})\,\overline{J}_2\,(Q_1 X_4^{-1})$ for the SEP (7.44). Hence

$$X_4 Q_1^{-1}\left(\alpha \overline{J}_3 - \beta \overline{J}_4\right) Q_1 X_4^{-1} z = \lambda z \tag{7.54}$$

which gives

$$\left(\alpha \underbrace{X_4 Q_1^{-1} \overline{J}_3 Q_1 X_4^{-1}}_{\tilde{J}_3} - \beta \underbrace{X_4 Q_1^{-1} X_3}_{\tilde{J}_4}\right) z = \lambda z \tag{7.55}$$

Remark 7 According to Sylvester's Law of inertia, the similar matrices J_2, $\overline{J}_2$ and $\tilde{J}_2$ have the same inertia, i.e. same number of positive, negative and null eigenvalues.

Theorem 6 *The equilibrium point* $(i_{Lie}, V_{ie}, \sigma_{ie}, e_{ie})$ *of the reduced system* (7.37)–(7.38) *is exponentially stable if* (7.45) *is satisfied and*

$$\beta > \frac{\sum_{i=1}^{n} \frac{1}{C_i R_i}}{\min_{y^T y \neq 0} \frac{y^T \Lambda_2 y}{y^T y}} \forall i \in \mathcal{I}. \tag{7.56}$$

where Λ_2 *is diagonal having the same index of inertia as matrix* $\overline{J}_4$.

Proof Notice that matrix X_3 is semi-positive definite with kernel spanned by $\left[\mathbf{0}_n^T \ \mathbf{1}_n^T\right]^T$ corresponding to the global synchronisation of the graph, while matrix $X_4 Q_1^{-1}$ is positive definite with kernel spanned by $\mathbf{0}_{2n}$. Following the proof of Theorem 8 in [17], by applying the Courant-Fischer Theorem to the eigenvalue problem, for global synchronisation of the graph, all eigenvalues of $\tilde{J}_4$ are real and negative since $Im\left(\mathcal{L}\right) = \mathbf{1}_n^{\perp}$, and $Im\left(X_4 Q_1^{-1}\right) \cap ker\left(X_3\right) = \mathbf{0}_{2n}$, which means that $X_3 z$ is never in the kernel of $X_4 Q_1^{-1}$. Hence, one can see that $ker\left(\tilde{J}_4\right) = ker\left(X_3\right)$. As the image of the matrix $X_4 Q_1^{-1}$ excludes $span\left(\left[\mathbf{0}_n^T \ \mathbf{1}_n^T\right]^T\right)$, it follows that $\tilde{J}_4 z$ is the zero vector if and only if $z \in span\left(\left[\mathbf{0}_n^T \ \mathbf{1}_n^T\right]^T\right)$ that corresponds to the global synchronisation of the graph.

Matrices $\overline{J}_3$ and $\tilde{J}_3$ have the same spectrum, that is, they have $2n - 1$ null eigenvalues and one positive eigenvalue. It is important to underline that multiplying $\tilde{J}_3$ with vector $z = [\mathbf{0}_1 \ \mathbf{1}_n]$ would render a value outside of the matrix spectrum. Hence, one can conclude that $z = \left[\mathbf{0}_n^T \ \mathbf{1}_n^T\right]^T$ is not an eigenvector of $\tilde{J}_3$, and hence, not an eigenvector of $\tilde{J}_2$. That is, vector $z = [\mathbf{0}_1 \ \mathbf{1}_n]$ does not belong in the eigenspace of the Jacobian matrix $\tilde{J}_2$, i.e. $z \notin N\left(\tilde{J}_2 - \lambda I_{2n}\right)$. A comprehensive explanation is given below.

Consider matrix $\overline{J}_3$ for which there is

$$\overline{J}_3 y = \lambda y \tag{7.57}$$

with λ the eigenvalue of $\overline{J}_3$, and y the corresponding eigenvector. Following the similarity transformation in (7.54), there is $\tilde{J}_3 z = \lambda z$, with z given as

$$z = \left(X_4 Q_1^{-1}\right)^{-1} y \tag{7.58}$$

Matrix $X_4 Q_1^{-1}$ is given in Sect. 7.3.3, and its inverse, $\left(X_4 Q_1^{-1}\right)^{-1}$, has the following expression

$$\left(X_4 Q_1^{-1}\right)^{-1} = \frac{1}{|\,\mathbb{L}\,|}\begin{bmatrix} D^{-1} & D^{-1} \\ D^{-1} & [V_e]^{-1}\left(V^* I_n + [e]\right) D^{-1} + [V_e]\, m R^{-1} D^{-1} \end{bmatrix} \tag{7.59}$$

where $\mathbb{L}$ is expressed in (7.52). Notice that (7.59) is also symmetric and positive-definite since $X_4 Q_1^{-1} \succ 0$, according to Proposition 2. For z given as $z = \left[\mathbf{0}_n^T \ \mathbf{1}_n^T\right]^T$, then there is

$$\left(X_4 Q_1^{-1}\right)^{-1}\begin{bmatrix} y_{n1} \\ y_{n2} \end{bmatrix} = \begin{bmatrix} \mathbf{0}_n \\ \mathbf{1}_n \end{bmatrix} = \begin{bmatrix} \frac{1}{|\mathbb{L}|}\left(D^{-1}\right)\left(y_{n1} + y_{n2}\right) \\ \frac{1}{|\mathbb{L}|}\left(D^{-1} y_{n1} + [V_e]^{-1}\left(V^* I_n + [e]\right) D^{-1} + [V_e]\, m R^{-1} D^{-1} y_{n2}\right) \end{bmatrix} \tag{7.60}$$

which holds if $y_{n1} = -y_{n2}$. However, if $y = [y_{n1} \; - y_{n1}]^T$, then from Eq. (7.57), eigenvector y would correspond to a positive eigenvalue outside of the spectrum of matrix $\overline{J}_3$. Thus, y is not an eigenvector of $\overline{J}_3$ since the opposite implies that $\overline{J}_3$ has two non-zero eigenvalues, which would be in contradiction with the proof of Proposition 1. Therein, it is demonstrated that the index of inertia of $\overline{J}_3$ is $i\left(\overline{J}_3\right) = [1 \;\; 0 \;\; 2n-1]$.

Therefore, it follows from Lemma 1, that if $z = \left[\mathbf{0}_n^T \; \mathbf{1}_n^T\right]^T$ is not an eigenvector of matrix $\tilde{J}_3$, then ultimately it is not an eigenvector of the Jacobian $\tilde{J}_2$.

Note that matrix $J_5 = \left(P_1^{-1} Q_1 X_4^{-1}\right) \tilde{J}_2 \left(X_4 Q_1^{-1} P_1\right)$, isospectral with $\tilde{J}_2$, can be expressed as

$$J_5 = \alpha \Lambda_1 - \beta P_1^{-1} X_3 X_4 Q_1^{-1} P_1 \tag{7.61}$$

Then, let the matrix $J_6 = P_3^{-1} J_5 P_3$, isospectral with J_5 and $\tilde{J}_2$, with $P_3 = P_1^{-1} P_2$, unitary according to Lemma 2, such that the new standard eigenvalue problem becomes

$$\left(\alpha P_3^{-1} \Lambda_1 P_3 - \beta \Lambda_2\right) w = \lambda w. \tag{7.62}$$

Thus, matrix $P_3^{-1} \Lambda_1 P_3$ is symmetric, since P_3 is unitary, similar to Λ_1, having the same index of inertia as $X_1 X_2$ as shown in Proposition 1, and Λ_2 diagonal having the same index of inertia as $X_3 X_4 Q_1^{-1}$, as explained in Proposition 2. Lemma 1 can be applied for the eigenvalues problem as follows

$$\alpha \sum_{i=1}^{n} \frac{1}{\alpha C_i R_i} - \beta \min_{w^T w \neq 0} \frac{w^T \Lambda_2 w}{w^T w} < 0. \tag{7.63}$$

$$\sum_{i=1}^{n} \frac{1}{C_i R_i} - \beta \min_{w^T w \neq 0} \frac{w^T \Lambda_2 w}{w^T w} < 0. \tag{7.64}$$

The above condition is satisfied at all times, with a proper choice of the gain β, required to satisfy condition (7.56).

Hence, if condition (7.56) holds, matrix J_6 is Hurwitz, and by similarity both J_5 and $\tilde{J}_2$ are also Hurwitz, with the latter having the same index of inertia as $\overline{J}_2$ and J_2 (as already mentioned in Remark 7). Then, there exist $\rho_2 > 0$ and a domain $\tilde{D}_x = \left\{x \in R^{2n}, \| x \|_2 < \rho_2\right\}$ where $\tilde{D}_x \subseteq D_x$ such that the reduced model is exponentially stable at the origin. This completes the proof. □

According to Theorem 4, there exists $\varepsilon^* = min\{\frac{min\{L_i\}}{r_v^*}, \frac{1}{k^*} > 0\}$ such that for all $\varepsilon < \varepsilon^*$ (or equivalently $\frac{r_{vi}}{L_i} > \frac{r_v^*}{min\{L_i\}}$, or $k_i > k^*$), the equilibrium point

$\left[i_{Le}^T \; V_e^T \; \sigma_e^T \; e_e^T\right]^T$ of (7.29)–(7.31) with $\sigma_{ie} \in (-\frac{\pi}{2}, \frac{\pi}{2})$ is exponentially stable; thus completing the stability analysis of the entire DC microgrid.

7.4 Simulation Results

To test the theoretical findings, simulations are performed in MATLAB/Simulink, using the average model of the bidirectional DC/DC boost converter. The DC microgrid considered for testing is presented in Fig. 7.3, with the parameters specified in Table 7.1, selected to guarantee the stability analysis, and with the maximum current i^{max} chosen to satisfy $i^{max} = E_{max}/r_v$. There are five converters and one common CPL. All components are connected through communication links, with all links assumed bidirectional to feature a balanced Laplacian matrix. Each source is driven by a bidirectional boost converter connected in parallel to a common CPL. The reference voltage is set to $V^* = 400\,\text{V}$. We assume that the ratio of the output power among the sources is $5:4:3:2:1$. According to Fig. 7.3, only converters 1 and 5 participate in the load voltage recovery, that is, the pinning control gains are set as $g_1 = g_5 = 1$ and $g_2 = g_3 = g_4 = 0$.

During the first 4 s, the load power demand is $P = 2\,\text{kW}$ and the system is controlled by the primary controller only. The load voltage is kept below the reference V^*, having $V_o \approx 393\,\text{V}$ as depicted in Fig. 7.4b. Also, it is clear from Fig. 7.4c, that the power sharing is not accurate, since the output currents are not proportional, i.e. $i \approx [1.6 \;\; 1.22 \;\; 1.15 \;\; 0.7266 \;\; 0.4]\,\text{A}$.

At $t = 4\,\text{s}$, the secondary controller is enabled, enhancing the performance of the system with the load voltage tightly regulated to the reference, $V_o = V^* = 400\,\text{V}$ (Fig. 7.4b), and the power sharing becoming very accurate with proportional output currents being $i \approx [1.66 \;\; 1.33 \;\; 1 \;\; 0.66 \;\; 0.33]\,\text{A}$ as one can notice in Fig. 7.4c, given

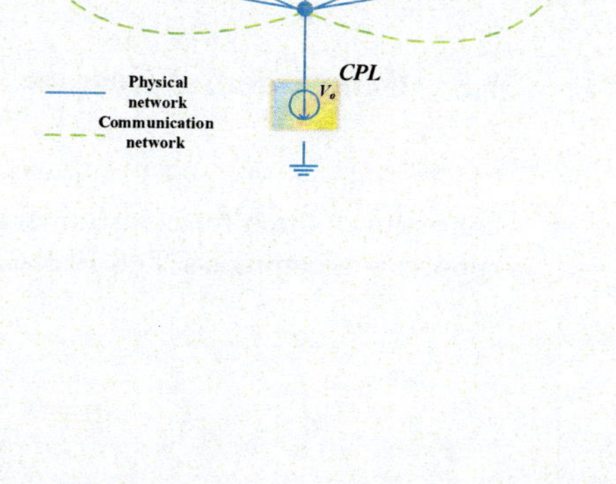

Fig. 7.3 DC microgrid prototype considered for simulation testing consisting of five converters, each of them communicating with their respective neighbours, with Converter 1 and Converter 5 sampling the load voltage and participating in the voltage restoration

Table 7.1 System and control parameters of a DC microgrid consisting of five bidirectional DC/DC boost converters feeding a common CPL

Parameters	Values
U [V]	[200 150 250 100 240]
C [μF]	[700 400 500 100 150]
R [Ω]	[0.5 1.5 1 0.7 1.2]
L [mH]	[2.3 2.2 2 2 2.5]
k	[1.8 2 2.5 1 1.9]
m	[0.014 0.0105 0.084 0.0420 0.0210]
E_{max}	[25 35 32 18 24]
r_v	[5 5 4 3 2]
α	100
β	10

the inductor currents being below their maximum technical limit as depicted in Fig. 7.4a.

The load power demand increases to $P = 3\,\text{kW}$, at $t = 8\,\text{s}$. In Fig. 7.4b, one can see that the load voltage remains at the desired 400 V, while the output currents are still accurately shared, in Fig. 7.4c, having proportional values $i \approx [2.5\ \ 2\ \ 1.5\ \ 1\ \ 0.5]\,\text{A}$.

In order to test the overcurrent protection, the system is required, at $t = 12\,\text{s}$, to feed an increased load of 4 kW. The load voltage stays fixed at 400 V (Fig. 7.4b). But, the inductor current of the second converter, i_{L2}, reaches its limit (Fig. 7.4a), and as a consequence, the converter loses its power sharing. Still, the power sharing is kept between the other four converters in a $5:3:2:1$ ratio (Fig. 7.4c), having $i \approx [3.35\ \ 2.02\ \ 1.35\ \ 0.67]\,\text{A}$.

7.4.1 Communication Failure

So as to investigate the influence of the possible communication failures on the performances of the proposed method, at $t = 16\,\text{s}$, the communication network is subjected to two faults. The links that connect converter 5 to the common bus is disconnected ($g_5 = 0$), and also the connection between converters 1 and 5 is disrupted. In Fig. 7.4b, c, one can observe that the voltage remains at the desired value V^* and the power sharing is unaffected by the communication failure.

7.5 Experimental Results

The DC microgrid setup displayed in Fig. 7.5 is considered for experimental testing, consisting of three Texas Instruments modules operated as DC/DC boost converters, supplying a common ETPS ELP-3362F electronic load acting as a CPL, with the

Fig. 7.4 Dynamic response of the DC microgrid system under primary and secondary controller

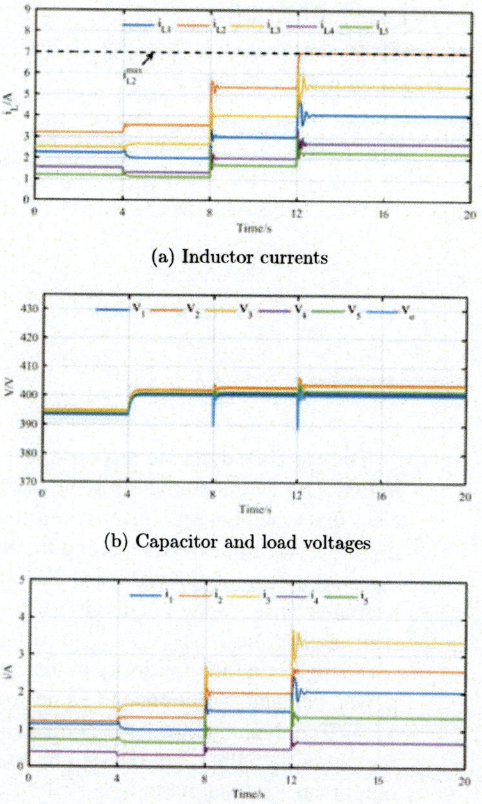

(a) Inductor currents

(b) Capacitor and load voltages

(c) Output currents

parameters given in Table 7.2. A more detailed electrical diagram of the testbed is presented in APPENDIX B. The main tasks of the primary and secondary controllers are to regulate the output voltage to $V^* = 48$ V, while keeping a proportional $1:1:2$ output load power sharing, provided none of the converters violate their maximum allowed input current, imposed by their technical requirements. The filtered dynamic response of the input/output currents and output voltages is presented in Fig. 7.6.

In Fig. 7.6a, under primary control only, the load power demand increases from 40 to 50 W. One can notice that the power sharing is not accurately kept in a $1:1:2$ ratio, having $i \approx [0.18\ \ 0.21\ \ 0.47]$ A when the load is 40 W and $i \approx [0.23\ \ 0.26\ \ 0.60]$ A when it increases to 50 W. Moreover, the load voltage regulation decreases from $V_o \approx 46.6$ V, as it was initially, down to $V_o \approx 45.9$ V at the steady state following the load change.

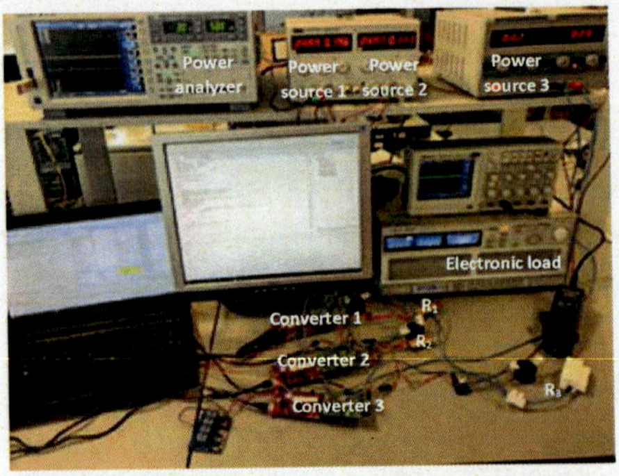

Fig. 7.5 Experimental testbed

The system's dynamic response when enabling the secondary control is captured in Fig. 7.6b, while maintaining the load power demand constant at 40 W. It becomes clear that when the secondary controller is enabled, the accuracy of the power sharing is visibly improved, reaching the desired proportional $1:1:2$ sharing, having the output currents $i \approx [0.22\ \ 0.22\ \ 0.44]$ A. Moreover, the load voltage rises and becomes closer to the rated value V^* compared to the case where only the primary control is applied.

To highlight the superiority of the combined primary and secondary control under power demand variation, the same load power change is performed, from 40 to 50 W. The dynamic response is presented in Fig. 7.6c. Unlike case (a), the voltage regulation is tighter, i.e. the output voltage V_o is closer to V^*, while the improved output currents maintain their $1:1:2$ desired sharing, with the output currents $i \approx [0.27\ \ 0.27\ \ 0.55]$ A.

Table 7.2 System and control parameters of a DC microgrid consisting of three Texas Instruments modules operated as DC/DC boost converters feeding an ETPS ELP-3362F electronic load acting as a CPL

Parameters	Values	Parameters	Values
U [V]	[24 24 24]	m	[3 6 6]$\times 10^{-3}$
C [μF]	[100 100 100]	E_{max}	[10 9 10]
R [Ω]	[1 1 2.3]	r_v	[5 6 5]
L [mH]	[2.2 2.2 2.2]	α	2
k	[5 5 10]	β	1

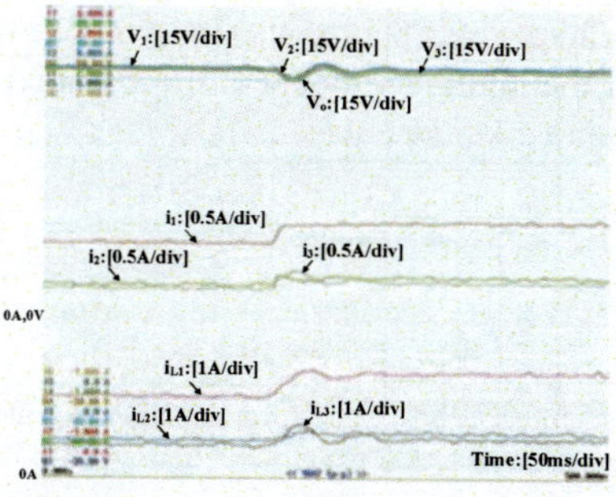

(a) Load power demand increases from 40 W to 50 W under primary control

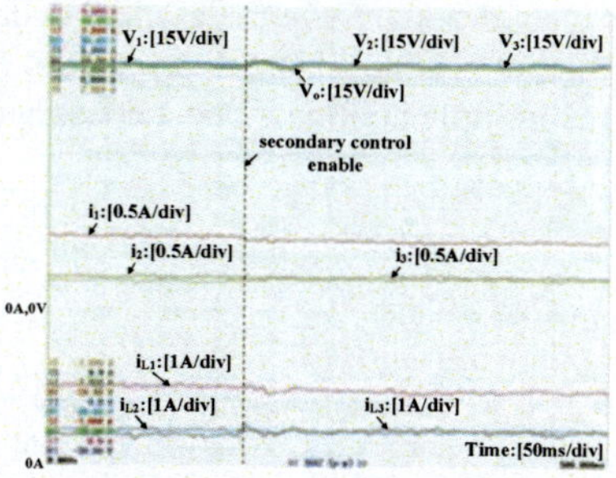

(b) Dynamic response when enabling the secondary control

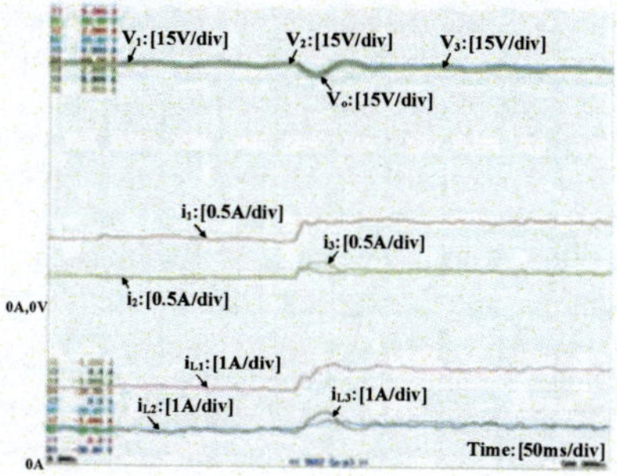

(c) Load power demand increases from 40 W to 50 W under primary and secondary control

Fig. 7.6 Experimental results of the DC microgrid system under primary and secondary controller

7.6 Conclusions

A novel decentralised primary and distributed secondary control was proposed to achieve accurate power sharing, voltage regulation, input current limitation and overcome the CPL instability problem. By employing singular perturbation theory and two time-scale analysis, the closed-loop system stability was analytically proven, taking into account both the physical system and the two-level control dynamics. Both simulation and experimental testings were carried out to validate the presented approach and analysis.

The impact of time delays, which may occur in the secondary control implementation, on the stability of the entire microgrid is of great interest (see [18–20]). Several methods for computing the maximum delay to avoid instability have been proposed, such as Pade approximations [21] or Rekasius substitution-based algorithm [22]. In the same framework, Lyapunov-based methods [23, 24], such as the Implicit Lyapunov Krasovski Functional (ILKF) have emerged to provide sufficient stability conditions. Nevertheless, the main aim of this chapter was to introduce for the first time this novel two-level control for the nonlinear model of the DC microgrid with multiple nonlinear boost converters and guarantee its stability.

References

1. G.C. Konstantopoulos, P.R. Baldivieso-Monasterios, State-limiting PID controller for a class of nonlinear systems with constant uncertainties. Int. J. Robust Nonlinear Control **30**(5), 1770–1787 (2019)
2. R. Ortega, A. Loria, P.J. Nicklasson, H. Sira-Ramirez, *Passivity-Based Control of Euler-Lagrange Systems: Mechanical, Electrical and Electromechanical Applications* (Springer, Berlin, 1998)
3. A.C. Braitor, G.C. Konstantopoulos, V. Kadirkamanathan, Current-limiting droop control design and stability analysis for paralleled boost converters in DC microgrids. IEEE Trans. Control Syst. Technol. **29**(1), 385–394 (2020)
4. A.-C. Braitor, G.C. Konstantopoulos, V. Kadirkamanathan, Admittance matrix computation and stability analysis of droop controlled dc micro-grids with CPL. IFAC-PapersOnLine **53**(2), 13525–13530 (2020)
5. A. Braitor, G.C. Konstantopoulos, V. Kadirkamanathan, Stability analysis and nonlinear current-limiting control design for DC micro-grids with CPLS. IET Smart Grid **3**(3), 355–366 (2020)
6. Z. Shuai, D. He, J. Fang, Z.J. Shen, C. Tu, J. Wang, Robust droop control of DC distribution networks. IET Renew. Power Gener. **10**(6), 807–814 (2016)
7. F. Cingoz, A. Elrayyah, Y. Sozer, Plug-and-play nonlinear droop construction scheme to optimize islanded microgrid operations. IEEE Trans. Power Electron. **32**(4), 2743–2756 (2017)
8. J.W. Simpson-Porco, F. Dörfler, F. Bullo, Voltage stabilization in microgrids via quadratic droop control. IEEE Trans. Autom. Control **62**(3), 1239–1253 (2017)
9. G. Xu, D. Sha, X. Liao, Decentralized inverse-droop control for input-series-output-parallel DC-DC converters. IEEE Trans. Power Electron. **30**(9), 4621–4625 (2015)
10. C. Jin, P. Wang, J. Xiao, Y. Tang, F.H. Choo, Implementation of hierarchical control in DC microgrids. IEEE Trans. Industr. Electron. **61**(8), 4032–4042 (2014)
11. J.-W. Kim, H.-S. Choi, B.H. Cho, A novel droop method for converter parallel operation. IEEE Trans. Power Electron. **17**(1), 25–32 (2002)
12. X. Lu, J.M. Guerrero, K. Sun, J.C. Vasquez, An improved droop control method for dc microgrids based on low bandwidth communication with DC bus voltage restoration and enhanced current sharing accuracy. IEEE Trans. Power Electron. **29**(4), 1800–1812 (2014)
13. P.H. Huang, P.C. Liu, W. Xiao, M.S. El Moursi, A novel droop-based average voltage sharing control strategy for DC microgrids. IEEE Trans. Smart Grid **6**(3), 1096–1106 (2015)
14. Z. Liu, M. Su, Y. Sun, H. Han, X. Hou, J.M. Guerrero, Stability analysis of DC microgrids with constant power load under distributed control methods. Automatica **90**, 62–72 (2018)
15. H.K. Khalil, *Nonlinear Systems*, 3rd edn. (Pearson, 2014)
16. M. Su, Z. Liu, Y. Sun, H. Han, X. Hou, Stability analysis and stabilization methods of DC microgrid with multiple parallel-connected DC-DC converters loaded by CPLs. IEEE Trans. Smart Grid **9**(1), 132–142 (2018)
17. J.W. Simpson-Porco, F. Dörfler, F. Bullo, Synchronization and power sharing for droop-controlled inverters in islanded microgrids. Automatica **49**(9), 2603–2611 (2013)
18. R. Olfati-Saber, R.M. Murray, Consensus problems in networks of agents with switching topology and time-delays. IEEE Trans. Autom. Control **49**(9), 1520–1533 (2004)
19. F. Milano, M. Anghel, Impact of time delays on power system stability. IEEE Trans. Circuits Syst. I Regul. Pap. **59**(4), 889–900 (2012)

20. E.A. Coelho, D. Wu, J.M. Guerrero, J.C. Vasquez, T. Dragicevic, C. Stefanovic, P. Popovski, Small-signal analysis of the microgrid secondary control considering a communication time delay. IEEE Trans. Industr. Electron. **63**(10), 6257–6269 (2016)
21. G.A. Baker Jr., P. Graves-Morris, *Pade Approximants* (Cambridge University Press, New York, 1996)
22. H. Jia, X. Cao, X. Yu, P. Zhang, A simple approach to determine power system delay margin, in *2007 IEEE Power Engineering Society General Meeting* (IEEE, New York, 2007), pp. 1–7
23. Y.L. Jiang, T. Jiang, C. Dong, A novel LMI criterion for power system stability with multiple time-delays. Sci. China Technol. Sci. **57**, 1392–1400 (2014)
24. A. Polyakov, D. Efimov, W. Perruquetti, J. Richard, Implicit Lyapunov-Krasovski functionals for stability analysis and control design of time-delay systems. IEEE Trans. Autom. Control **60**(12), 3344–3349 (2015)

Chapter 8
Droop-Controlled DC Microgrids with Overvoltage Protection

As the need for overvoltage protection has emerged recently, in this chapter, a droop controller with an inherent overvoltage protection will be developed at the primary layer of the hierarchical control framework. The proposed droop controller follows different designs, but it is implemented similarly both in parallel configuration and meshed configuration networks, as shown in the following sections.

In Sect. 8.1, the droop control strategy is introduced to limit the voltage of each paralleled source below an imposed limit, ensure tight voltage regulation and accurate load power distribution, and guarantee closed-loop system asymptotic stability in the presence of a common CPL. The upper bound of the voltage of each source is diligently proven using ultimate boundedness theory, while by using the admittance matrix factorisation of the microgrid developed in Chap. 4, analytic sufficient conditions for stability are acquired to lead the control parameters design. A detailed simulation scenario for a parallel configuration DC microgrid, having the converter units equipped with the developed controller, is presented to validate the theoretical design and analysis.

In Sect. 8.2, a slightly modified droop control methodology for meshed DC microgrids with CPLs is proposed, which guarantees the crucial overvoltage protection property of each DER unit, independently from each other or the loads. Following the acquisition of the admittance matrix, also known as loopy-Laplacian [1], of meshed DC microgrids, asymptotic stability to the desired equilibrium for the closed-loop system is analytically proven, rendering detailed stability conditions. Simulation testing is performed for a meshed DC microgrid to verify the theoretical contribution and the effectiveness of the proposed primary controller.

A.-C. Braitor, *Advanced Hierarchical Control and Stability Analysis of DC Microgrids*, Springer Theses, https://doi.org/10.1007/978-3-030-95415-4_8

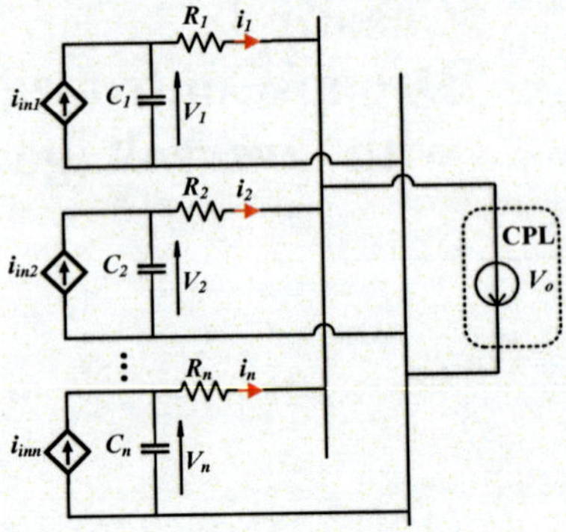

Fig. 8.1 Generic framework of a DC microgrid

8.1 Parallel Configuration DC Microgrid

Let us consider again the typical microgrid topology shown in Chap. 4 (Fig. 4.1), consisting of n power converters having an output capacitor, C_i, connected in parallel to a common DC bus through a line/cable with resistance, R_i, and supplying a common generic load. For the CPL case, the reduced model when assuming a fast inner current control loop becomes as shown in Fig. 8.1. The model can be described by the differential equations describing the capacitor voltages, acquired by employing Kirchhoff's laws, as follows:

$$C_i \dot{V}_i = i_{ini} - i_i \tag{8.1}$$

where V_i is the capacitor voltage, while i_{ini} and i_i represent the input and output current respectively, with i_{ini} also being used as the control input, for $\forall i \in \mathcal{I}$. One can notice that this framework represents but a generic model of n-sourced units which could be introduced in the microgrid structure via different power converter configurations (i.e. buck, boost, buck-boost, or AC/DC).

When a CPL is present, the power balance equation has to be satisfied:

$$P = V_o \sum_{i=1}^{n} i_i \tag{8.2}$$

with V_o being the voltage at the common bus, and P the load power.

Similar to Chap. 4, the output current i_i has the following expression

$$i_i = \frac{V_i - V_o}{R_i} \tag{8.3}$$

Let Assumption 1 hold. Replacing i_i from (8.3) into (8.2), it yields the following expression for the common bus voltage, obtained from the real solutions of a second order polynomial, similar to (4.7) as

$$V_o = \frac{\sum_{i=1}^{n} \frac{V_i}{R_i} \pm \sqrt{\left(\sum_{i=1}^{n} \frac{V_i}{R_i}\right)^2 - 4P\sum_{i=1}^{n} \frac{1}{R_i}}}{2\sum_{i=1}^{n} \frac{1}{R_i}}.$$

As already mentioned, the load voltage (4.7) has two solutions, a high voltage and a low voltage, with the high voltage being the feasible solution given as in (4.8).

When taking the partial derivative of the output current i_i from (8.3), with respect to the output voltage V_i, one obtains the admittance matrix Y [2], as in (4.11)

$$Y = R^{-1}\left(I_n - \mathbf{1}_{n\times n}D\right),$$

where $R = diag\{R_i\}$, and $D = diag\left\{\frac{\partial V_o}{\partial V_i}\right\} \succ 0$ from (4.12) having the following expression

$$D = \frac{1}{2\sum_{i=1}^{n} \frac{1}{R_i}} \left(R^{-1} + \frac{\sum_{i=1}^{n} \frac{V_i}{R_i}}{\sqrt{\left(\sum_{i=1}^{n} \frac{V_i}{R_i}\right)^2 - 4P\sum_{i=1}^{n} \frac{1}{R_i}}} R^{-1} \right)$$

where, according to Assumption 1, the denominator $\sqrt{\left(\sum_{i=1}^{n} \frac{V_i}{R_i}\right)^2 - 4P\sum_{i=1}^{n} \frac{1}{R_i}} > 0$. The eigenvalues of D will have the form

$$\lambda_{Di} = \frac{1}{2\sum_{i=1}^{n} \frac{1}{R_i}} \left(\frac{1}{R_i} + \frac{\sum_{i=1}^{n} \frac{V_i}{R_i}}{\sqrt{\left(\sum_{i=1}^{n} \frac{V_i}{R_i}\right)^2 - 4P\sum_{i=1}^{n} \frac{1}{R_i}}} \frac{1}{R_i} \right),$$

with $i \in \mathcal{I}$.

8.1.1 *Proposed Control Architecture*

The proposed control strategy is set to ensure tight voltage regulation close to the desired value, and accurate load power distribution among the DER units with an inherent overvoltage protection for each source independently from the system parameters and the load [3].

8.1.1.1 Droop Control Design with Overvoltage Protection

Motivated by the sl-PI controller proposed in [4], the novel droop-based method defines the control input i_{ini}, in the following way

$$i_{ini} = -g_i V_i + I_{maxi} sin\sigma_i \tag{8.4}$$

where σ is designed to follow the nonlinear dynamics

$$\dot{\sigma}_i = \frac{k_i}{I_{maxi}}(V^* - V_o - m_i i_i)cos\sigma_i \tag{8.5}$$

which incorporates the droop control, having

$$m_i < R_i. \tag{8.6}$$

By substituting the controller dynamics into the open-loop system, it yields

$$C_i \dot{V}_i = -g_i V_i + I_{maxi} sin\sigma_i - i_i. \tag{8.7}$$

Consider the following continuously differentiable energy-like function

$$W_i = \frac{1}{2} C_i V_i^2 \tag{8.8}$$

and by taking its time derivative, one obtains

$$\begin{aligned} \dot{W}_i &= -g_i V_i^2 + V_i I_{maxi} sin\sigma_i - V_i i_i \\ &= -g_i V_i^2 + V_i I_{maxi} sin\sigma_i - P_i, \end{aligned} \tag{8.9}$$

where $P_i = V_i i_i$ is the power injected by the ith source into the common bus through each line i. Depending on the positive/negative sign of the power P_i, the proof is divided into two cases:

(a) $P_i \geq 0$

From Eq. (8.9), it can be clearly observed that

$$\dot{W}_i \leq -g_i V_i^2 + V_i I_{maxi} sin\sigma_i \leq -g_i |V_i|^2 + I_{maxi} |V_i|. \tag{8.10}$$

Let $g_i = \bar{g}_i + \epsilon_i > 0$, with $\bar{g}_i > 0$, and ϵ_i being an arbitrarily small positive constant. Then, inequality (8.10) takes the form

$$\begin{aligned} \dot{W}_i &\leq -(\bar{g}_i + \epsilon_i) |V_i|^2 + I_{maxi} |V_i| \\ &\leq -\epsilon_i |V_i|^2, \ \forall |V_i| \geq \frac{I_{maxi}}{\bar{g}_i}. \end{aligned} \tag{8.11}$$

According to (8.11), the solution $V_i(t)$ is uniformly ultimately bounded. Moreover, every solution starting with the initial condition $V_i(0)$ that satisfied

$$|V_i(0)| \leq \frac{I_{maxi}}{\bar{g}_i}, \tag{8.12}$$

will stay in this range for all future times, as

$$|V_i(t)| \leq \frac{I_{maxi}}{\bar{g}_i}, \ \forall t \geq 0. \tag{8.13}$$

To guarantee that each voltage V_i is limited below a maximum value V^{max}, the control parameters $\bar{g}_i$ and I_{maxi} can be chosen to satisfy the equality

$$\frac{I_{maxi}}{\bar{g}_i} = V^{max}. \tag{8.14}$$

Hence, this concludes the design part of the control parameters $\bar{g}_i$ and I_{maxi}, to ensure an upper bound for the capacitor voltage V_i, when the power $P_i \geq 0$.

(b) $P_i < 0$

Given the DC microgrid structure and the existence of a constant power load, having $P > 0$, then at least one current source (e.g. jth source) should be feeding the CPL and/or other (up to $n-1$) sources. Thus, if the corresponding power of that particular source is $P_j > 0$, then since $P_j = V_j i_j = V_j \frac{V_j - V_o}{R_j}$, it yields that $V_j > V_o$, and equivalently from Case (a), there is $V_o < V_j \leq V^{max}$. However, since for the ith source, the output power is negative $P_i = V_i \frac{V_i - V_o}{R_i} < 0$, then $V_i < V_o$ which eventually leads to $V_i < V^{max}$.

Therefore, independently of the sign of the source units power, an upper bound for the capacitor voltage is ensured, i.e. $V_i(t) \leq V^{max}$, at any time instant, even during transients.

8.1.1.2 Parameter Selection

One should note that the control parameters I_{maxi} and $g_i = \bar{g}_i + \epsilon_i$ can take any values that satisfy equality (8.14) in order to ensure the required overvoltage protection. However, to provide a practical guidance for the user to choose these two values, a worst case scenario is considered where the ith source feeds the load by itself, i.e. $P_i \approx P$. Then, it can be easily understood from (8.9) that depending on the value of P, compared to the term $g_i V_i^2$, the actual upper bound of V_i can be limited well below V^{max}. Considering a known upper value of the CPL power, i.e. $0 < P \leq P_{max}$, and since it is desired that the upper value of V_i to be as close to V^{max} as possible, one can achieve this by suitably selecting the parameter g_i such that the term $g_i V_i^2$ dominates the term P in (8.9), i.e. it is at least 10 times higher assuming a worst case scenario, thus

$$g_i \geq \frac{10 P_{max}}{(V^{max})^2}. \tag{8.15}$$

Subsequently, since according to (8.14) there is $\bar{g}_i V^{max} = I_{maxi}$, then I_{maxi} can be then selected as

$$I_{maxi} \approx \frac{10 P_{max}}{V^{max}}, \tag{8.16}$$

due to the very small positive constant ϵ_i. Note that the above expressions for selecting the controller parameters are provided for guidance only, since any other selection that satisfies (8.14) will still guarantee the desired upper limit for each voltage V_i. Additionally, a more detailed analysis on the condition that the parameter g_i needs to satisfy is provided in the sequel and is related to the asymptotic stability of the closed-loop system.

8.1.2 Stability Analysis

Let the closed-loop system be written in a matrix form as

$$\dot{V} = C^{-1}(-gV + I_{max}\mathbf{sin}\sigma - i) \tag{8.17}$$

$$\dot{\sigma} = I_{max}^{-1} k\,[\mathbf{cos}\sigma]\left(\left(V^* - V_o\right)\mathbf{1}_n - mi\right) \tag{8.18}$$

where $C = diag\{C_i\}$, $V = [V_1 \dots V_n]^T$, $g = diag\{g_i\}$, $I_{max} = diag\{I_{maxi}\}$, $i = [i_1 \dots i_n]^T$, $\sigma = [\sigma_1 \dots \sigma_n]^T$, $k = diag\{k_i\}$, $m = diag\{m_i\}$.

Considering an equilibrium point (V_e, σ_e) of the closed-loop system (8.17)–(8.18), (8.3) and (4.7), with $\sigma_{ie} = \left(-\frac{\pi}{2}, \frac{\pi}{2}\right)$, that satisfies Assumption 1, the following theorem can be formulated that guarantees stability of the entire droop-controlled DC microgrid with a CPL.

Theorem 7 *The equilibrium point (V_e, σ_e) is asymptotically stable if the controller parameter g_i satisfies*

$$g_i > \frac{n\lambda_{Di} - 1}{R_i}, \forall i \in \mathcal{I}. \tag{8.19}$$

Proof The corresponding Jacobian matrix of system (8.17)–(8.18) has the following form

$$J = \begin{bmatrix} -C^{-1}g - C^{-1}Y & C^{-1}I_{max}\,[\mathbf{cos}\sigma_e] \\ -I_{max}^{-1}k\,[\mathbf{cos}\sigma_e]\,(\mathbf{1}_{n\times n}I_n + mY) & \mathbf{0}_{n\times n} \end{bmatrix} \tag{8.20}$$

Replacing Y from (4.11), one gets

$$J = \begin{bmatrix} -C^{-1}\left(g - R^{-1}\left(I_n - \mathbf{1}_{n\times n}D\right)\right) & C^{-1}I_{max}\,[\mathbf{cos}\sigma_e] \\ -I_{max}^{-1}k\,[\mathbf{cos}\sigma_e]\left(\mathbf{1}_{n\times n}D + mR^{-1}\left(I_n - \mathbf{1}_{n\times n}D\right)\right) & \mathbf{0}_{n\times n} \end{bmatrix}, \tag{8.21}$$

with the characteristic polynomial of the system, according to Theorem 1, given as

$$|\lambda I_{2n} - J| = |\lambda^2 I_n + \lambda\mathbb{C} + \mathbb{K}| = 0, \tag{8.22}$$

where the two matrix coefficients are

$$\mathbb{C} = C^{-1}g + C^{-1}R^{-1}\left(I_n - \mathbf{1}_{n\times n}D\right)$$

$$\mathbb{K} = C^{-1}\left[\mathbf{cos}\sigma_e\right]^2 k\left(\left(I_n - mR^{-1}\right)\mathbf{1}_{n\times n}D + mR^{-1}\right).$$

By right multiplication with $|D^{-1}| > 0$, Eq. (8.22) becomes

$$|\lambda^2 D^{-1} + \lambda\bar{\mathbb{C}} + \bar{\mathbb{K}}| = 0 \tag{8.23}$$

with

$$\bar{\mathbb{C}} = C^{-1}gD^{-1} + C^{-1}R^{-1}\left(D^{-1} - \mathbf{1}_{n\times n}\right)$$

$$\bar{\mathbb{K}} = C^{-1}\left[\mathbf{cos}\sigma_e\right]^2 k\left(\left(I_n - mR^{-1}\right)\mathbf{1}_{n\times n} + mR^{-1}D^{-1}\right)$$

By left multiplying (8.23) with $|RC| > 0$, it yields

$$|\lambda^2 RCD^{-1} + \lambda\mathbb{C}^* + \mathbb{K}^*| = 0 \tag{8.24}$$

with

$$\mathbb{C}^* = RgD^{-1} + D^{-1} - \mathbf{1}_{n\times n}$$

$$\mathbb{K}^* = R\left[\mathbf{cos}\sigma_e\right]^2 k\left(I_n - mR^{-1}\right)\times$$
$$\left(\mathbf{1}_{n\times n} + \left(I_n - mR^{-1}\right)^{-1} k^{-1}\left[cos\sigma\right]^{-1} R^{-1}mD^{-1}\right)$$

Notice that matrix $\mathbb{C}^*$ is a symmetric matrix, and, after factorisation, matrix $\mathbb{K}^*$, according to Lemma 4, is a diagonalisable matrix with real eigenvalues. Thus, by expressing the latter as $\mathbb{K}^* = P^{-1}\Lambda P$, with matrix P being orthogonal, and Λ diagonal, and replacing it in (8.24), it yields that

$$|\lambda^2 RCD^{-1} + \lambda\mathbb{C}^* + P^{-1}\Lambda P| = 0 \tag{8.25}$$

or, equivalently,

$$|\lambda^2 PRCD^{-1}P^{-1} + \lambda P\mathbb{C}^*P^{-1} + \Lambda| = 0 \tag{8.26}$$

which is a quadratic eigenvalue problem (QEP) with matrix Λ diagonal, having the eigenvalues of matrix $\mathbb{K}^*$ on the main diagonal, whereas the similarity transformations $PRCD^{-1}P^{-1}$ and $P\mathbb{C}^*P^{-1}$ are symmetrical, since P is orthogonal $\left(P^{-1} = P^T\right)$, and they have the same eigenvalues as matrices RCD^{-1}, and $\mathbb{C}^*$, respectively. According to the QEP theory (Lemma 5), if RCD^{-1}, $\mathbb{C}^*$, and Λ are positive-definite, then the eigenvalues λ will be real and negative, i.e. $\lambda < 0$, thus, J will be Hurwitz.

Since $RCD^{-1} \succ 0$, the remaining conditions are $\mathbb{C}^* \succ 0$, and $\Lambda \succ 0$ (or equivalently $\mathbb{K}^*$ has positive eigenvalues).

1. $\mathbb{C}^* \succ 0$: As $\mathbb{C}^*$ is a sum of symmetric matrices, according to Lemma 1, the condition becomes
$$\frac{R_i g_i + 1}{\lambda_{Di}} - n > 0$$
which is always guaranteed, provided (8.19) holds.
2. $\Lambda \succ 0$ (or equivalently, $\mathbb{K}^*$ has positive eigenvalues): Due to the choice in (8.6), and for the bounded $\sigma_{ie} \in \left(-\frac{\pi}{2}, \frac{\pi}{2}\right)$, the first matrix term in the multiplication inside $\mathbb{K}^*$ is positive-definite, i.e.
$$R\,[\mathbf{cos}\sigma_e]^2\, k\left(I_n - mR^{-1}\right) \succ 0.$$
Therefore, according to Lemma 2, one can investigate only the remaining symmetrical matrix in the product, which is
$$\mathbf{1}_{n\times n} + \left(I_n - mR^{-1}\right)^{-1} k^{-1}\,[\mathbf{cos}\sigma_e]^{-1}\, R^{-1} m D^{-1} \succ 0.$$
The above matrix is represented by a sum between a positive semi-definite and a positive-definite symmetric matrices, hence, one can clearly agree that the matrix is positive-definite (Lemma 1).

As a result, when (8.19) is satisfied, J is Hurwitz, and the equilibrium point (V_e, σ_e) is asymptotically stable. This completes the proof. □

8.1.3 Simulation Results

A DC microgrid portrayed in Fig. 8.2, with the parameters specified in Table 8.1, chosen according to Sect. 8.1.1.2, consisting of five DC/DC buck converters, is simulated in MATLAB/Simulink, considering the average model of the buck converter for a 0.3 s testing scenario. The desired task for the proposed controller is to regulate the load voltage close to the reference value, $V^* = 100$ V, and accurately distribute the load power among converters in a 5 : 4 : 3 : 2 : 1 ratio, while maintaining a safe output voltage margin below 1.05 V*, i.e. 5% above the rated value.

The simulation starts at $t = 0$ s, with a load power demand being $P = 250$ W. As one can notice in the time responses in Fig. 8.3, the load voltage is tightly regulated very close to the reference with $V_o \approx 99.95$ V, as expected by the droop control function (Fig. 8.3b). The output currents are accurately shared in a 5 : 4 : 3 : 2 : 1 ratio with $i = [0.66\ 0.53\ 0.4\ 0.26\ 0.13]$, as it can be seen in Fig. 8.3a. Note that the output voltages are kept below their upper limit.

Later on, at $t = 0.1$ s, the load power demand increases to $P = 1$ kW. Notice that the transient occurring shortly after the load change, at 0.1 s, is successfully limited

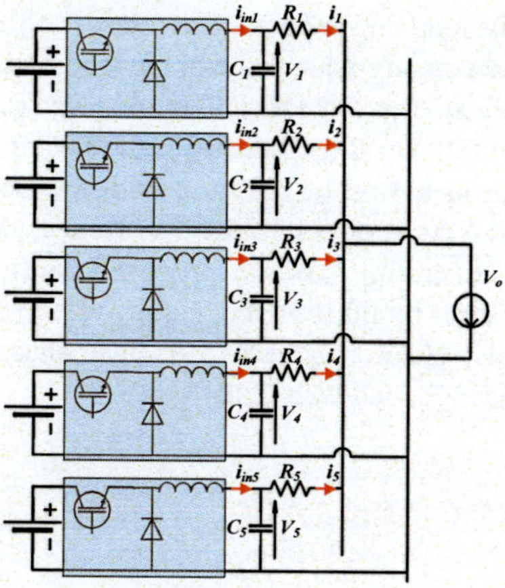

Fig. 8.2 DC microgrid considered for testing

Table 8.1 System and control parameters of a DC microgrid consisting of five DC/DC buck converters feeding a common CPL

System parameters	Values
$C_{1...5}$ [μF]	[25 50 20 20 5]
$R_{1...5}$ [Ω]	[1 1.1 1.05 1.12 1.5]
Control parameters	Values
$m_{1...5}$	[0.42 0.21 0.14 0.105 0.084]
$I_{max1...5}$	[1.05 1.05 1.05 1.05 1.05] $\times$ 6000
$g_{1...5}$	[1 1 1 1 1] $\times$ 60
$k_{1...5}$	[2 2 2 2 38] $\times$ 10^7

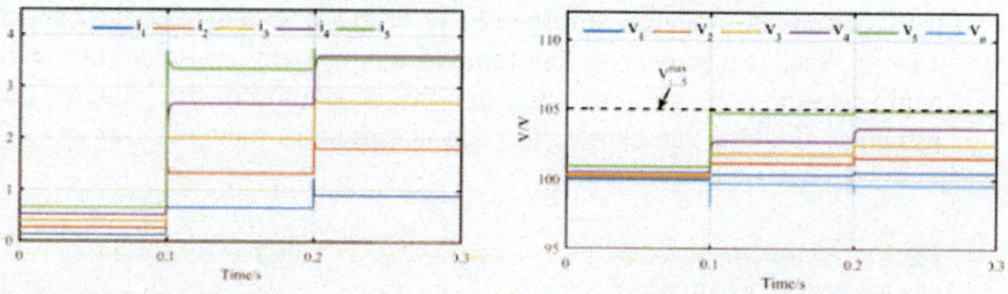

(a) Output currents (b) Output voltages

Fig. 8.3 Simulation results of the DC microgrid equipped with the proposed controller

below 105 V. As one can see in Fig. 8.3b, the load voltage is still kept very close to the rated with $V_o \approx 99.7$ V, and the power sharing is also very accurate with the output current vector being $i = [3.34\ 2.67\ 2\ 1.34\ 0.67]$, as displayed in Fig. 8.3a.

At $t = 0.2$ s, the load power demand increases further to $P = 1.25$ kW. According to Fig. 8.3b, the new steady-state value of the load voltage is $V_o = 99.6$ V, meanwhile the output voltage of the 5th DC/DC buck converter is limited below the upper bound $V^{max} = 105$ V. On the other hand, the power sharing among the other four converters is kept in a $4:3:2:1$ ratio with the output current vector being $i = [3.6\ 2.7\ 1.8\ 0.9]$ A as shown in Fig. 8.3a. Hence, the theoretic analysis has been clearly verified, illustrating how the proposed controller has as its first priority to protect each converter by limiting the output voltage below a desired upper bound at all times, while also ensures the required power sharing and load voltage regulation in the DC microgrid loaded by a CPL.

8.2 Meshed DC Microgrid

8.2.1 DC Microgrid Model

A common meshed DC microgrid architecture is depicted in Fig. 8.4, consisting of a finite number of nodes n, each of the nodes representing a controllable DER unit supplying a local CPL and connected with each other through resistive lines. Note that if a different microgrid architecture was considered, where some nodes include a load but not a DER source, the system can still be transformed into the one in Fig. 8.1 using the Kron-reduced network approach [1]. In Fig. 8.5, the model of the voltage source converter that integrates each DER unit with each node j is depicted. The dynamic equations of the capacitor voltages for a random node j can be acquired by employing Kirchhoff's laws

$$C_j \dot{V}_j = i_{inj} - i_j \tag{8.27}$$

where C_j is the output capacitor, V_j is the output voltage, while i_{inj} and i_j represent the input and output current respectively, with i_{inj} also used as the control input, for $\forall j \in \mathcal{I}$. This is a typical system representation where an inner current controller is applied to the converter, resulting in a fast regulation of the inductor current to the value i_{inj} [5]. One can express the output current of the converter as

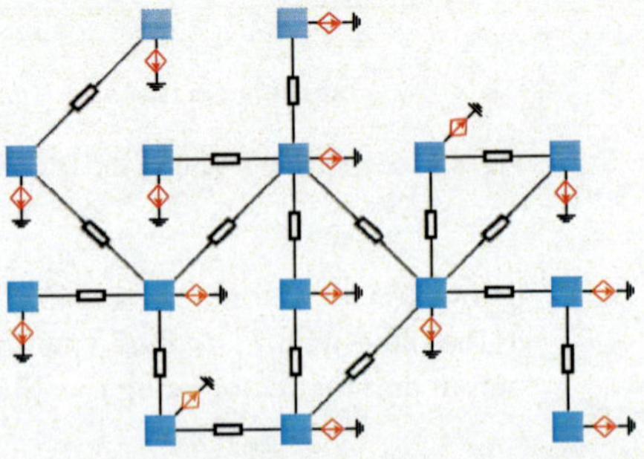

Fig. 8.4 Generic framework of a meshed DC microgrid

Fig. 8.5 Integration of a DER unit in a meshed DC microgrid through a power converter

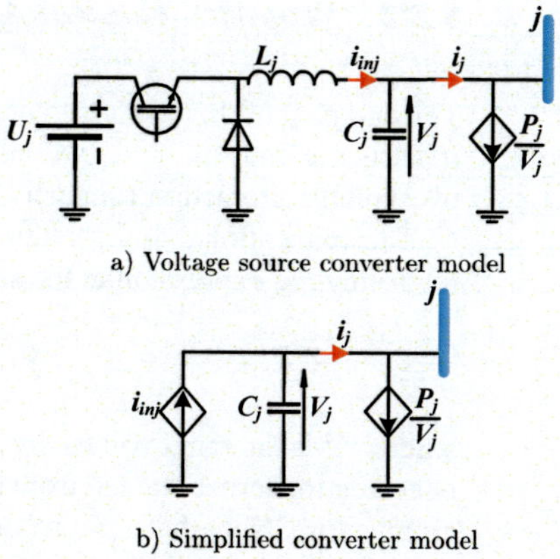

$$i_j = \frac{P_j}{V_j} + \sum_{k\in\mathcal{N}} i_{jk}. \tag{8.28}$$

where $\mathcal{N}$ represents the neighbourhood of the node j, in the induced graph $\mathcal{G}$ described by the meshed DC network, i.e. $\mathcal{N} \in \mathcal{V} : \varepsilon_{jk} \in \mathcal{E}$.

Remark 8 As mentioned in Remark 4, one can notice that the system configuration (8.27) represents just a generic model of n-sourced units which can be introduced within the microgrid structure via different power converter configurations (buck, boost, buck-boost, or AC/DC), where a fast inner current control loop is considered.

Considering a steady-state voltage value for the jth node denoted by V_{je}, by taking the partial derivative of the output current i_j from (8.28) with respect to the output voltage V_j, one can obtain the symmetric admittance matrix Y of the DC microgrid, in the following form:

$$Y = \begin{bmatrix} -\frac{P_1}{V_{1e}^2} + \sum_{k\in\mathcal{N}} \frac{1}{R_{1k}} & -\frac{1}{R_{12}} & \cdots & -\frac{1}{R_{1n}} \\ -\frac{1}{R_{12}} & -\frac{P_2}{V_{2e}^2} + \sum_{k\in\mathcal{N}} \frac{1}{R_{2k}} & \cdots & -\frac{1}{R_{2n}} \\ \vdots & \vdots & \ddots & \vdots \\ -\frac{1}{R_{1n}} & -\frac{1}{R_{2n}} & \cdots & -\frac{P_n}{V_{ne}^2} + \sum_{k\in\mathcal{N}} \frac{1}{R_{nk}} \end{bmatrix}. \tag{8.29}$$

If there is no connection between the vertices j and k, i.e. $\varepsilon_{jk} \notin \mathcal{E}$, the corresponding Y matrix entry will be zero, i.e. $\frac{1}{R_{jk}} = 0$. The admittance matrix Y can be rewritten as

$$Y = L - D, \tag{8.30}$$

with $D = diag\left\{\frac{P_j}{V_{je}^2}\right\}$ positive-definite and L positive-semidefinite matrices. Note that L represents the Laplacian matrix of the graph $\mathcal{G}$ induced by the DC microgrid, while D incorporates the self-loops of the nodes.

8.2.2 Proposed Control Architecture

The end goal of this work is to design a primary controller for the DC microgrid that inherits the conventional and widely used droop controller, whilst guaranteeing an overvoltage protection for each DER unit (node) independently [6].

The conventional droop control approach requires each node voltage V_j to satisfy the following expression at the steady-state:

$$V_j = V^* - m_j i_j + x_j^{set} \tag{8.31}$$

where V^* is the rated voltage, m_j is the droop coefficient and x_j^{set} is a desired signal or correction term obtained from the supervisory controller in the hierarchical control architecture. Note that x_j^{set} can be set to 0, which is a common approach in islanded DC microgrids [7], but generally it can represent a constant or piecewise constant value due to the time-scale separation difference between primary and supervisory control.

This section investigates only the primary control dynamics, and in order to achieve the desired goal, as mentioned above, a novel primary droop control technique is proposed in the sequel.

8.2.2.1 Droop Control Design with Overvoltage Protection

Motivated by the development of the sl-PI controller in [4], the proposed droop control strategy defines the control input, i_{inj}, in the following manner

$$i_{inj} = -g_j V_j + I_{maxj} sin\sigma_j \tag{8.32}$$

with σ constructed to follow the nonlinear dynamics

$$\dot{\sigma}_j = \frac{k_j}{I_{maxj}}(V^* - V_j - m_j i_j + x_j^{set}) cos\sigma_j \tag{8.33}$$

that incorporates the droop control, with the droop coefficient m_j satisfying the following inequality

$$m_j < 1. \tag{8.34}$$

Substituting the controller dynamics (8.32)–(8.33) into the open-loop system (8.27), it yields

$$C_j \dot{V}_j = -g_j V_j + I_{maxj} sin\sigma_j - i_j. \tag{8.35}$$

By taking the following continuously differentiable energy-like function for each node j

$$W_j = \frac{1}{2} C_j V_j^2 \tag{8.36}$$

and by calculating its time derivative, it becomes

$$\dot{W}_j = -g_j V_j^2 + V_j I_{maxj} sin\sigma_j - V_j \left(\frac{P_j}{V_j} + \sum_{k \in \mathcal{N}} i_{jk} \right)$$
$$= -g_j V_j^2 + V_j I_{maxj} sin\sigma_j - \left(P_j + V_j \sum_{k \in \mathcal{N}} i_{jk} \right), \tag{8.37}$$

where $V_j \sum_{k \in \mathcal{N}} i_{jk}$ represents the power fed by the jth converter to the neighbouring converters through every ε_{jk} edge. Considering that $V_j \sum_{k \in \mathcal{N}} i_{jk}$ could be both positive or negative, this leads to the scenario where similarly the total power $P_j + V_j \sum_{k \in \mathcal{N}} i_{jk}$ could be positive or negative. Hence, the boundedness of the voltage V_j is not straightforward. As a result, the proof can be divided into the two following distinct cases:

(a) Case 1: $P_j + V_j \sum_{k \in \mathcal{N}} i_{jk} \geq 0$
From Eq. (8.37), it is clear that

$$\dot{W}_j \leq -g_j V_j^2 + V_j I_{maxj} sin\sigma_j \leq -g_j |V_j|^2 + I_{maxj} |V_j|. \tag{8.38}$$

Let $g_j = \bar{g}_j + \epsilon_j > 0$, with $\bar{g}_j > 0$ and ϵ_j representing an arbitrarily small positive constant. In that case, (8.38) becomes

$$\dot{W}_j \leq -\left(\bar{g}_j + \epsilon_j\right) |V_j|^2 + I_{maxj} |V_j|$$
$$\leq -\epsilon_j |V_j|^2, \ \forall |V_j| \geq \frac{I_{maxj}}{\bar{g}_j}. \tag{8.39}$$

According to (8.39), the solution $V_j(t)$ is uniformly ultimately bounded, and every solution starting with the initial condition $V_j(0)$, satisfying

$$|V_j(0)| \leq \frac{I_{maxj}}{\bar{g}_j}, \tag{8.40}$$

will remain in this range for all future time, i.e.

$$|V_j(t)| \leq \frac{I_{maxj}}{\bar{g}_j}, \ \forall t \geq 0. \tag{8.41}$$

To ensure that each voltage V_j is bounded below a maximum voltage V^{max}, the control parameters, $\bar{g}_j$ and I_{maxj} can be selected to satisfy

$$\frac{I_{maxj}}{\bar{g}_j} = V^{max}. \tag{8.42}$$

This completes the design of the control parameters $\bar{g}_j$ and I_{maxj}, to guarantee an upper bound for the output voltage V_j, when $P_j + V_j \sum_{k\in\mathcal{N}} i_{jk} \geq 0$.

(b) Case 2: $P_j + V_j \sum_{k\in\mathcal{N}} i_{jk} < 0$
In the islanded microgrid case, considering the existence of constant power loads with $P_j > 0$, at least one converter (e.g. kth converter) should be feeding the loads and/or other (up to $n-1$) converter units, based on Kirchhoff's laws. Hence, if the corresponding power of that particular source is $P_k + V_k \sum_{l\in\mathcal{N}} i_{kl} > 0$, then since $P_k + V_k \sum_{l\in\mathcal{N}} i_{kl} = V_k \left(\frac{P_k}{V_k} + \sum_{l\in\mathcal{N}} i_{kl}\right)$, it yields that $V_k > V_l$, and equivalently from Case 1, there is $V_k \leq V^{max}$. However, since for the jth source, the output power is negative $P_j < 0$, then there always exists a spanning tree in the induced connected graph $\mathcal{G}$ such that $V_j < V_l < V_k$ which leads to $V_j < V^{max}$.

Therefore, in both cases, an upper bound for the output voltage is guaranteed, i.e. $V_j(t) \leq V^{max}$, at any time instant, i.e. even during transients.

8.2.3 *Stability Analysis*

Consider now the closed-loop microgrid system written in matrix form:

$$\dot{V} = C^{-1}\left(-gV + I_{max}\mathbf{sin}\sigma - i\right) \tag{8.43}$$

$$\dot{\sigma} = I_{max}^{-1} k\left[\mathbf{cos}\sigma\right]\left(V^*\mathbf{1}_n - V - mi\right) \tag{8.44}$$

where $C = diag\left\{C_j\right\}$, $V = [V_1 \dots V_n]$, $g = diag\left\{g_j\right\}$, $I_{max} = diag\left\{I_{maxj}\right\}$, $i = [i_1 \dots i_n]$, $\sigma = [\sigma_1 \dots \sigma_n]$, $k = diag\left\{k_j\right\}$, $m = diag\left\{m_j\right\}$. Considering an equilibrium point (V_e, σ_e) of the closed-loop system (8.43)–(8.44), with $\sigma_{ie} = \left(-\frac{\pi}{2}, \frac{\pi}{2}\right)$, the following theorem can be formulated that guarantees stability of the entire droop-controlled DC microgrid with a CPL.

Theorem 8 *The equilibrium point (V_e, σ_e) is asymptotically stable if the controller parameter g_j satisfies*

$$g_j > \frac{P_j}{V_{je}^2}, \quad \forall j \in \mathcal{I}. \tag{8.45}$$

Proof The corresponding Jacobian matrix of system (8.43)–(8.44) has the following form

$$J = \begin{bmatrix} -C^{-1}g - C^{-1}Y & C^{-1}I_{max}\left[\mathbf{cos}\sigma_e\right] \\ I_{max}^{-1}k\left[\mathbf{cos}\sigma_e\right](I_n + mY) & \mathbf{0}_{n\times n} \end{bmatrix} \tag{8.46}$$

Replacing the admittance matrix Y with its expression from (8.30), it yields

$$J = \begin{bmatrix} -C^{-1}g - C^{-1}(L-D) & C^{-1}I_{max}\,[\mathbf{cos}\sigma_e] \\ I_{max}^{-1}k\,[\mathbf{cos}\sigma_e]\,(I_n + m\,(L-D)) & \mathbf{0}_{n\times n} \end{bmatrix} \tag{8.47}$$

According to Theorem 1, the characteristic polynomial will look as follows

$$|\lambda I_{2n} - J| = |\lambda^2 I_n + \lambda\mathbf{C} + \mathbf{K}| = 0 \tag{8.48}$$

with

$$\mathbf{C} = C^{-1}(g + L - D) \tag{8.49}$$

$$\mathbf{K} = C^{-1}[\mathbf{cos}\sigma_e]^2\,k\,(mL + I_n - mD) \tag{8.50}$$

By left multiplying (8.48) with $|m^{-1}k^{-1}[\mathbf{cos}\sigma_e]^{-2}\,C| > 0$, one obtains

$$|\lambda^2 m^{-1}k^{-1}[\mathbf{cos}\sigma_e]^{-2}\,C + \lambda\overline{\mathbf{C}} + \overline{\mathbf{K}}| = 0 \tag{8.51}$$

with

$$\overline{\mathbf{C}} = m^{-1}k^{-1}[\mathbf{cos}\sigma_e]^{-2}(g + L - D) \tag{8.52}$$

$$\overline{\mathbf{K}} = \left(L + m^{-1} - D\right) \tag{8.53}$$

Notice that matrix $\overline{\mathbf{K}}$ is symmetric and matrix $\overline{\mathbf{C}}$ diagonalisable according to Lemma 4, with $P^{-1}\overline{\mathbf{C}}P = \Lambda$, with P unitary and Λ diagonal, having the same index of inertia as matrix $\overline{\mathbf{C}}$. Equation (8.51) becomes

$$|\lambda^2 m^{-1}k^{-1}[\mathbf{cos}\sigma_e]^{-2}\,C + \lambda P^{-1}\Lambda P + \overline{\mathbf{K}}| = 0 \tag{8.54}$$

or equivalently

$$|\lambda^2 P m^{-1}k^{-1}[\mathbf{cos}\sigma_e]^{-2}\,C P^{-1} + \lambda\Lambda + P\overline{\mathbf{K}}P^{-1}| = 0 \tag{8.55}$$

which is a quadratic eigenvalue problem (QEP) with Λ diagonal having the same index of inertia as matrix $\overline{\mathbf{C}}$, and the similarity transformation $Pm^{-1}k^{-1}[\mathbf{cos}\sigma_e]^{-2}\,CP^{-1}$ and $P\overline{\mathbf{K}}P^{-1}$ symmetrical since P is unitary ($P^{-1} = P^T$), and isospectral with $m^{-1}k^{-1}[\mathbf{cos}\sigma_e]^{-2}\,C$ and $\overline{\mathbf{K}}$, respectively. According to the QEP theory presented in Lemma 5, if the matrix coefficients are positive-definite, then the eigenvalues are negative, i.e. $\lambda < 0$, thus the Jacobian is Hurwitz. Matrix $m^{-1}k^{-1}[\mathbf{cos}\sigma_e]^{-2}\,C$ is already positive definite, hence the two remaining conditions are:

1. $\Lambda \succ 0$, or equivalently $\overline{\mathbf{C}}$ has positive eigenvalues. Since matrix $\overline{\mathbf{C}}$ is represented by a product of two symmetric matrices, one of them being positive-definite, i.e. $m^{-1}k^{-1}[\mathbf{cos}\sigma_e]^{-2} \succ 0$, according to Sylvester's law of inertia, one can investigate the sign of the remaining symmetric matrix,

$$g + L - D \succ 0. \tag{8.56}$$

In the worst case scenario, by employing Lemma 1, the above condition becomes in scalar form

$$g_j + 0 - \frac{P_j}{V_{je}^2} > 0 \tag{8.57}$$

which holds true provided that (8.45) is satisfied.

2. $\overline{\mathbf{K}} \succ 0$ that, in the worst case scenario according to Lemma 1, in scalar form becomes

$$0 + \frac{1}{m_j} - \frac{P_j}{V_{je}^2} > 0 \tag{8.58}$$

which is true given the appropriate selection of the droop coefficient as specified in (8.34), given the current is generally smaller in network systems.

This completes the proof. □

8.2.4 Simulation Results

A DC microgrid portrayed in Fig. 8.6, with the parameters specified in Table 8.2, is considered for simulation testing, consisting of 7 DER units integrated via buck converters, each of them connected to a local CPL. The main objective of the proposed controller is to regulate each node voltage close to $V^* = 100$ V based on the droop controller concept, while guaranteeing an overvoltage protection.

The system dynamic response is presented in Fig. 8.7. During the first 0.02 s the converters operate in conventional droop control mode, as the constant correction term $x^{set} = [0\ 0\ 0\ 0\ 0\ 0\ 0]$. Prior to the first load change at 0.01 s, the load power demand is $P = [5\ 2\ 1\ 0.5\ 3\ 0.2\ 3] \times 10^2$ W. The output voltages drop just below

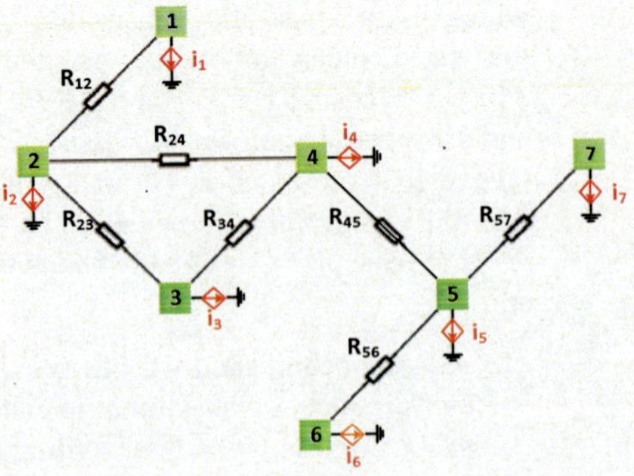

Fig. 8.6 Meshed DC microgrid under investigation

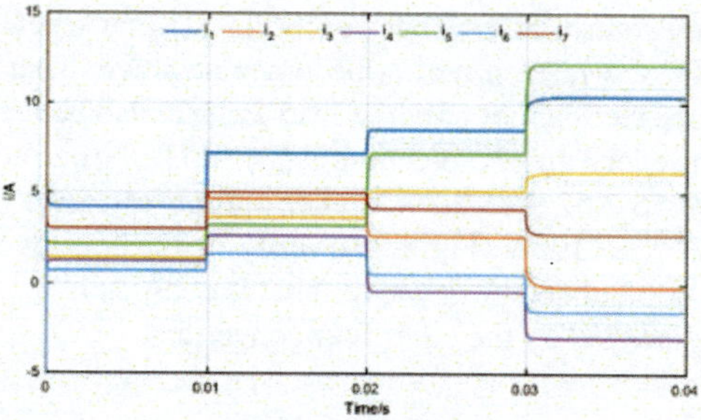

(a) Output currents

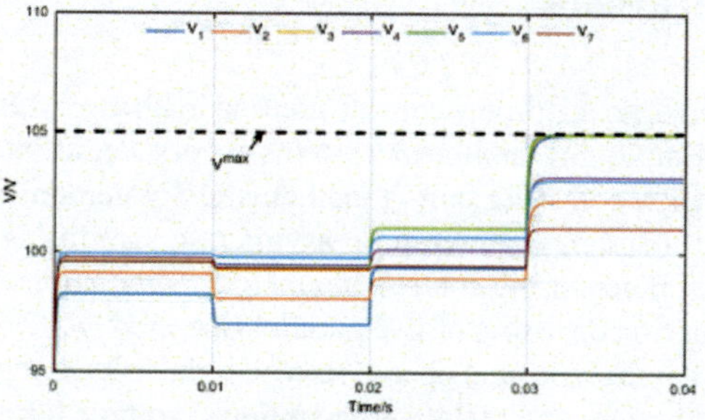

(b) Output voltages

Fig. 8.7 Simulation results of the DC microgrid equipped with the proposed controller

Table 8.2 System and control parameters of a meshed DC microgrid consisting of seven DC/DC buck converters each unit feeding a local CPL

Parameters	Values
$C\ [\mu F]$	[250 50 200 75 100 350 150]
m	[0.42 0.42 0.21 0.21 0.21 0.14 0.14]
I_{max}	$105 \times 2 \times 10^2$
g	2×10^2
k	2×10^7
$[R_{12}\ R_{23}\ R_{24}\ R_{34}\ R_{45}\ R_{56}\ R_{57}]\ [\Omega]$	[1 1.5 2 1.25 0.5 0.75 1.75]

the rated value of 100 V (Fig. 8.7b), as expected by the droop control feature, and the output currents are all positive, thus they all feed their local loads, as it can be observed in Fig. 8.7a.

At $t = 0.01$ s, the load power demand increases to $P = [8\ 5\ 3\ 1.5\ 4\ 1\ 5] \times 10^2$ W. From Fig. 8.7b, it can be noticed that the output voltages drop even lower than before, while the output currents increase to satisfy the new power demand as reported in Fig. 8.7a.

While maintaining the power demand constant, at $t = 0.02$ s, the correction term becomes $x^{set} = [2.94\ 0\ 2.1\ 0\ 2.52\ 0.7\ 0]$, representing possible input signals from a supervisory controller. One can see in Fig. 8.7 that several voltages increase above the rated 100 V, while current i_4 becomes negative. That means that the other six converters are feeding not only the load P_4, but also converter 4.

To test the overvoltage protection, at $t = 0.03$ s the correction term becomes $x^{set} =$ [10.08 2.1 7.56 2.52 9.45 2.8 1.4]. The output voltages $V_3,\ V_5,\ V_6$ are successfully limited to $V^{max} = 105$ V (Fig. 8.7b), verifying the developed theory, while the output currents $i_2,\ i_4,\ i_6$ become negative. That is converters 2, 4, 6 and their local loads, $P_2,\ P_4,\ P_6$, are fed by the other four converters.

8.3 Conclusions

In Sect. 8.1, a novel droop control method with overvoltage protection has been proposed. Employing nonlinear systems theory, an ultimate bound for the output voltage of each converter unit is mathematically demonstrated. Following the proposed control strategy, closed-loop asymptotic stability is also ensured pending an appropriate choice of the control parameters. Simulations have been carried out for a DC microgrid consisting of five parallel-operated DC/DC buck converters feeding a CPL, displaying a normal operation with tight voltage regulation and accurate load power distribution, and maintaining an upper voltage limit at all times, even during transient periods.

In Sect. 8.2, an enhanced droop controller with voltage limitation has been introduced for meshed DC microgrids consisting of multiple DER units and CPLs. Utilising nonlinear systems theory, an ultimate bound for the output voltage of each converter unit was rigorously proven. Closed-loop stability was rigorously guaranteed given some straightforward conditions are met. The theoretical findings and the effectiveness of the proposed approach were verified through simulation testing. The end goal of this approach was to present for the first time this novel primary droop control structure.

References

1. F. Dorfler, F. Bullo, Kron reduction of graphs with applications to electrical networks. IEEE Trans. Circuits Syst. I: Regul. Pap. **60**(1), 150–163 (2013)
2. A.-C. Braitor, G.C. Konstantopoulos, V. Kadirkamanathan, Admittance matrix computation and stability analysis of droop controlled DC micro-grids with CPL. IFAC-PapersOnLine **53**(2), 13525–13530 (2020). 21st IFAC World Congress
3. A.-C. Braitor, G.C. Konstantopoulos, V. Kadirkamanathan, Novel droop control design for overvoltage protection of DC microgrids with a constant power load, in *2020 28th Mediterranean Conference on Control and Automation (MED)* (2020), pp. 875–880
4. G.C. Konstantopoulos, P.R. Baldivieso-Monasterios, State-limiting PID controller for a class of nonlinear systems with constant uncertainties. Int. J. Robust Nonlinear Control n/a(n/a) (2019)
5. F. Gao, R. Kang, J. Cao, T. Yang, Primary and secondary control in DC microgrids: a review. J. Mod. Power Syst. Clean Energy **7**(2), 227–242 (2019)

6. A.-C. Braitor, G.C. Konstantopoulos, V. Kadirkamanathan, Enhanced primary droop controller for meshed DC micro-grids with overvoltage protection, in *2021 29th Mediterranean Conference on Control and Automation (MED)* (2021), pp. 368–373
7. A.C. Braitor, G.C. Konstantopoulos, V. Kadirkamanathan, Current-limiting droop control design and stability analysis for paralleled boost converters in DC microgrids. IEEE Trans. Control Syst. Technol. 1–10 (2020)

Chapter 9
Conclusions

A brief summary is assembled in this chapter, which puts a spotlight on the main development goals achieved in this thesis. It discusses a series of remarks and includes potential limitations and considered assumptions in this work. Additionally, it stresses the opportunity of extending the current work by providing future potential steps to improve and add into the existing contributions presented in the thesis.

9.1 Summary

To sum up, the main parts in this thesis are the following.

A novel idea to compute the admittance matrix is proposed in Chap. 4. Isolating singular matrices and suitably factorising the admittance matrix leads to a simpler path towards acquiring sufficient stability conditions. The end result has been utilised further in the stability analysis presented in Chaps. 5–7.

A current-limiting droop controller to ensure power sharing is developed in Chap. 5 for a DC microgrid consisting of unidirectional boost converters supplying a nonlinear Z, I or P load. Moreover, the controller also implements an inherent current limitation for each converter unit. Closed-loop stability of the overall DC microgrid and control system is analytically proven, and simulation and experimental results are presented to support the developed strategy by comparison to the conventional droop control.

The DC microgrid, in Chap. 6, incorporates bidirectional DC/DC boost converters and three-phase AC/DC rectifiers feeding a CPL connected at the main bus. The concept of incorporating a constant virtual resistance and a bounded dynamic virtual voltage ensures that the input currents of every converter unit will never violate the imposed maximum, regardless of the system parameters and without requiring any limiters or saturators. In doing so, the integrator wind-up and instability issues, that occur with traditional current-limiting designs, are avoided. Simulation testing

A.-C. Braitor, *Advanced Hierarchical Control and Stability Analysis of DC Microgrids*, Springer Theses, https://doi.org/10.1007/978-3-030-95415-4_9

scenarios are included with load power variations and charging/discharging battery cycles. In addition, a parallel configuration microgrid with an electronic load is tested experimentally.

In Chap. 7, by making use of the recently proposed state-limiting PI, a hierarchical control scheme is deployed to achieve accurate power sharing, voltage regulation, input current limitation and overcome the CPL instability problem. Applying the singular perturbation theory and two time-scale analysis, closed-loop system stability is guaranteed for the overall system incorporating the physical system and the multi-level control dynamics. Both simulation and experimental results are included to validate the presented strategy and analysis.

An enhanced droop control methodology with an overvoltage protection has been introduced in Chap. 8, for parallel and meshed configuration DC microgrids consisting multiple DER units and CPLs. By effectively employing nonlinear systems theory, an ultimate bound for the voltage of each source is mathematically demonstrated. Asymptotic stability was diligently investigated and ensured pending straightforward sufficient conditions. Simulation testing is carried out to verify the proposed approaches.

9.1.1 Discussions

One might notice that throughout this thesis the power lines/cables within the DC microgrids have been considered, for simplicity, as purely resistive in the stability theory part. That is because the aim was to prove the stability of the overall system, and as shown in [1], the line inductance would not have any effect on the system stability.

In Chap. 5, the input voltage of each converter is assumed constant and used in the control design, which might be an issue if the voltage would vary. But, a constant voltage might not be realistic in a DC microgrid application where the sources might be represented by a PV or a Li-ion battery. This shortcoming has been addressed later on, in Chap. 6, where the controller no longer uses the constant input voltage, but a virtual dynamic voltage appropriately defined in the control design phase.

Although, in this work, the network topologies revolved around parallel and meshed configuration, it can be agreed that any topology could reduce to one of the two, by using the Kron-reduced network approach [2].

9.1.2 Assumptions and Limitations

This research work has taken into account a series of assumptions, which at first might seem potentially restrictive. In fact, they are sensible premises that allow a clear theoretical analysis pursuit. The aim is to exclude any side discussions, which is why they are necessary and represent a good starting point, for instance:

- The inequalities introduced in Assumptions 1 and 6 are needed to ensure the existence of a high load voltage solution. If the power is high, the inequalities could be easily satisfied by just increasing the rated voltage in the DC microgrid. Similar assumptions have been taken into consideration in [3];
- Assumption 2, stating that the input voltage is less or equal to the output voltage, is guaranteed through the boost converter design;
- Following a straightforward mathematical proof, one can ensure based on the previously mentioned assumptions, that the inequalities found in Assumptions 3 and 7 are always satisfied;
- Since the proof of existence of a unique equilibrium point for a microgrid under primary (and secondary) control is a non-trivial problem (see [4, 5]), one is required to include Assumptions 4, 8, and 10;
- Inequalities introduced in Assumptions 5 and 9 depend on control parameters, thus, they can be guaranteed by incorporating a suitable selection of control parameters in the control design part that leads to inequalities being satisfied at all times.

But turning the focus beyond the big picture, found within these assumptions, there are particular cases that one could choose to dive into, such as investigating both high and low voltage solutions of the common bus. However, that was not the purpose of this thesis, but it could be a potential future step, one of many as explained in more detail below.

9.2 Future Work

The work carried out in this thesis could be further extended and the possibilities and research directions are numerous. Here are a few examples:

- ☐ In this work, one uses DC microgrids that have converters mostly in a parallel or meshed configuration, but other topologies might be interesting to study as well, for instance converters in a cascaded topology;
- ☐ The voltage regulation and power sharing results could be improved in Chaps. 5 and 6 by incorporating a secondary control, similar or different to the one presented in Chap. 7;
- ☐ The transient performance of the proposed controllers in Chaps. 5 and 6 can be refined to reduce oscillations, and incorporate the method within a hierarchical control scheme;
- ☐ One could investigate the effects onto the microgrid performance when time-delays in the measurement or the control implementation occur, under different combination of a series-parallel network;
- ☐ Similarly, in presence of a communication network, as the one introduced in Chap. 7, the consequences of time-delays onto system stability would be an interesting phenomenon to study;

□ Not so different to overvoltages in Chap. 8, undervoltages cause unwanted effects as well, and although not presented in this thesis, an undervoltage protection can be guaranteed using a modified version of the proposed controller in Chap. 8;
□ In Chap. 8, Sect. 8.2, an optimisation algorithm could be implemented further for the secondary control state x^{set} computing the optimal flow to reduce power losses between nodes in the meshed network. In doing so, x^{set} will not be a constant anymore.

References

1. A.P.N. Tahim, D.J. Pagano, E. Lenz, V. Stramosk, Modeling and stability analysis of islanded DC microgrids under droop control. IEEE Trans. Power Electron. **30**(8), 4597–4607 (2015)
2. F. Dorfler, F. Bullo, Kron reduction of graphs with applications to electrical networks. IEEE Trans. Circuits Syst. I: Regul. Pap. **60**(1), 150–163 (2013)
3. M. Su, Z. Liu, Y. Sun, H. Han, X. Hou, Stability analysis and stabilization methods of DC microgrid with multiple parallel-connected DC-DC converters loaded by CPLs. IEEE Trans. Smart Grid **9**(1), 132–142 (2018)
4. Z. Liu, M. Su, Y. Sun, H. Han, X. Hou, J.M. Guerrero, Stability analysis of DC microgrids with constant power load under distributed control methods. Automatica **90**, 62–72 (2018)
5. J.W. Simpson-Porco, F. Dörfler, F. Bullo, Voltage stabilization in microgrids via quadratic droop control. IEEE Trans. Autom. Control **62**(3), 1239–1253 (2017)

Appendix A
Schematic of the Texas Instruments DC/DC Boost Converter

Two identical Texas Instruments DC/DC boost converters (TI DCDC-Boost-2phs-HV [1]) in parallel, interfacing different DC sources, have been utilised in the experimental testing in both Sects. 5.2 and 6.2 feeding an ETPS ELP-3362F electronic load operating in constant power mode. The electrical schematic of the boost converter is displayed on the next page (Fig. A.1), and it also incorporates an interleaved circuit which was not used in this work.

This converter model was already equipped with input and output voltage sensors, as well as input current sensors. However, since both controllers in Sects. 5.2 and 6.2 require the value of the load voltage V_o, an additional voltage divider and extra wiring have been put into place to have access to the measurement of the common bus voltage (i.e. V_{b-out}).

Reference

1. TI DCDC-Boost 2phs HV, https://www.ti.com/lit/ug/tidu403/tidu403.pdf?ts

A.-C. Braitor, *Advanced Hierarchical Control and Stability Analysis of DC Microgrids*, Springer Theses, https://doi.org/10.1007/978-3-030-95415-4

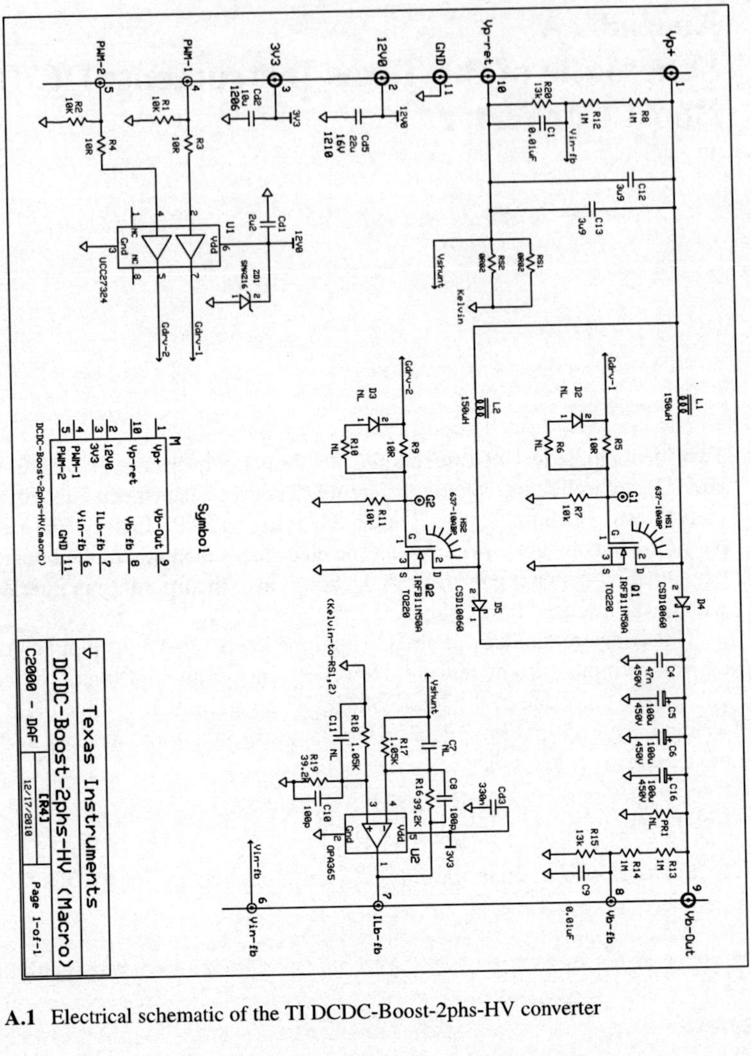

Fig. A.1 Electrical schematic of the TI DCDC-Boost-2phs-HV converter

Appendix B
Schematic of the Texas Instruments 3-Phase Inverters

In this case, three identical Texas Instruments three-phase inverters (TI BOOSTXL-3PhGaNInv [1]) in parallel, feeding an ETPS ELP-3326F electronic load operating in constant power mode, have been considered. However, only the lower switch of the first leg was used to simulate the operation of a DC/DC boost converter, with only one phase of the three-phase inverter being live. All the other switches are left open. The electrical diagram of the testbed is presented in Fig. B.1.

Additional voltage sensors have been put into place to measure the output voltages needed by the primary control layer. And for the secondary control layer, to ensure the information exchange within the units, cables between analog input/output ports have been employed to send and receive real-time power data used in the power sharing. Also, since the secondary controller needs the load voltage measurement as well, an extra voltage sensor was introduced to sample the load voltage and send it to the secondary controller corresponding to converter 1.

Reference

1. TI BOOSTXL-3PhGaNInv, https://www.ti.com/tool/boostxl-3phganinv

A.-C. Braitor, *Advanced Hierarchical Control and Stability Analysis of DC Microgrids*, Springer Theses, https://doi.org/10.1007/978-3-030-95415-4

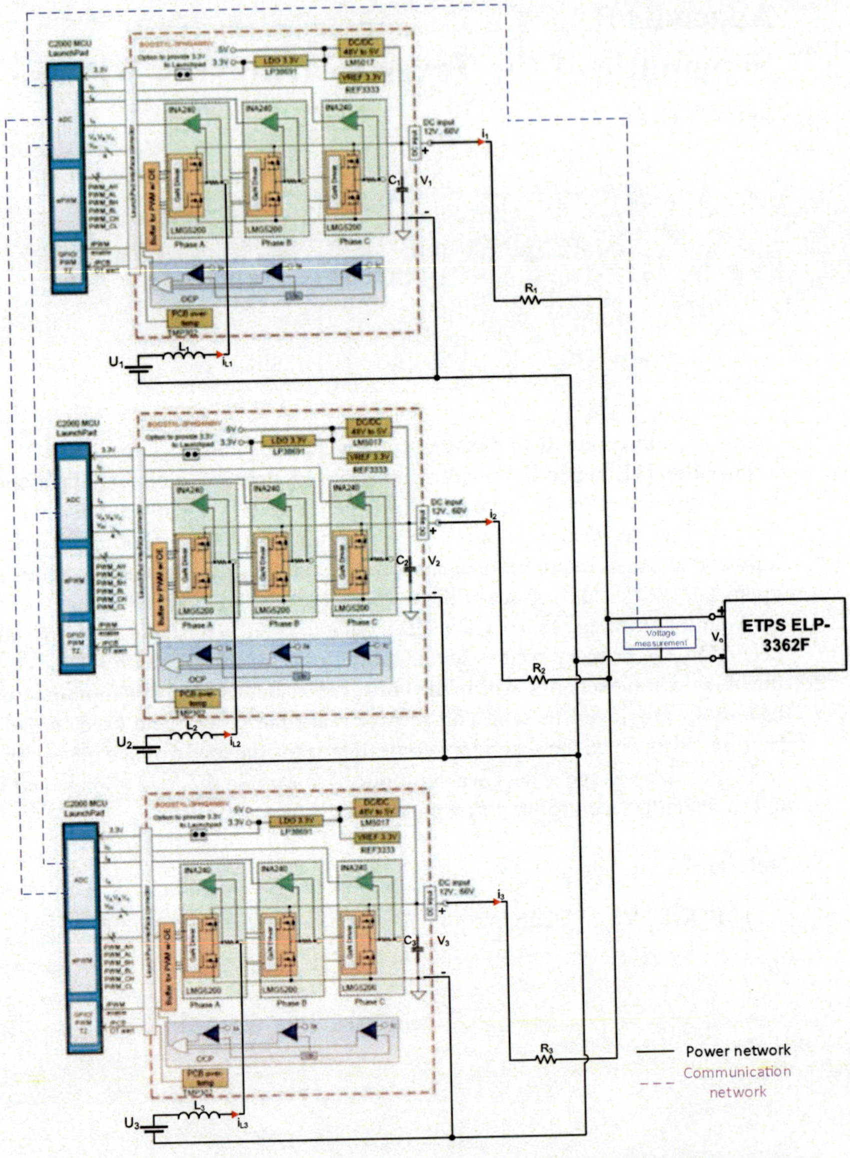

Fig. B.1 Experimental setup diagram including the 3 TI BOOSTXL-3PhGaNInv converters